Color Atlas of Freshwater Algae

Barry H. Rosen

Color Atlas of Freshwater Algae

Comprehensive Identification Guide, Including Harmful Algal Blooms

Springer

Barry H. Rosen
Florida Gulf Coast University
Ft. Myers, FL, USA

ISBN 978-3-031-93121-5 ISBN 978-3-031-93122-2 (eBook)
https://doi.org/10.1007/978-3-031-93122-2

This Springer imprint is published by the registered company Springer Nature Switzerland AG
The registered company address is: Gewerbestrasse 11, 6330 Cham, Switzerland

Preface

Microscopic observations of the inhabitants in aquatic environments reveal the complex array of organisms woven together that are informative, ornate, colorful, and artistic. The diversity of life, especially among the photosynthetic organisms called algae, displays a tapestry of patterns and structures that allow scientific names to be assigned. This guide is a compilation of over 1300 microphotographs of 340 genera that illustrate the beauty of algae and assist in their identification. The images in this guide initially started as an academic, printed, hands-on teaching guide based on the flora from Florida Gulf Coast University and then expanded to include a myriad of samples from across the United States. This guide organizes and arrays the images with morphological similarity to instruct a new generation of phycologists, those that study algae.

The identification and classification of algae from Linnaeus in 1767, through the nineteenth and twentieth centuries, mostly relied on pen and ink line drawings. These were carefully constructed to illustrate morphological features that separated species and are still a valuable resource. Both scanning electron microscopy (SEM) and transmission electron microscopy (TEM) brought significantly higher resolution than light microscopy, illustrating the external and internal structures through black-and-white imagery. As digital photography improved for the light microscope, color images of the key features that distinguish one organism from another have been generated.

Most of the images in this guide are light micrographs of live algae using differential interference contrast, an illumination technique that enhances physical relief. In addition, some of the images were created by a Z-stack: several images taken as the objective lens moves from the top of an object to the bottom and then assembled into an extended depth of focus single image. The silica cell wall of diatoms is best viewed after the cellular contents are digested (cleaned). Both cleaned and live-light microscope microphotographs of diatoms and some SEM and TEM images are included. I apologize for the images that are not crisp and clear; with time and better samples, I will endeavor to revise and improve this first edition. I acknowledge that misidentifications happen, and I look forward to suggestions for taxonomic revisions.

The relationship between organisms and their closest "relatives"-phylogeny has advanced significantly as genetic information has become commonly used. Many of the taxa presented in this morphological guide have names that have been revised based on new phylogenetic studies. The source of names used in the guide relied heavily on "AlgaeBase, a Listing of the World's Algae" (https://www.algaebase.org), a comprehensive and current source of the taxonomic status of the algae, along with some illustrations. In addition, AlgaeBase provides links to download many taxonomic works in their entirety. With diatoms being the exception, this guide relied on the "Diatoms of North America" (https://diatoms.org) for taxonomic position and status.

Abbreviations: cf. = "compare"; um=micron; sp.= species (singular); spp.=species (plural); var.=variety.

Ft. Myers, FL, USA Barry H. Rosen

Acknowledgments

My study of algae began under the direction of Dr. Francis R. Trainor at the University of Connecticut. His undergraduate, graduate, and independent study courses in phycology sparked my lifelong interest in algae. I was truly fortunate to find a Ph.D. advisor, mentor, and friend with Dr. Rex L. Lowe at Bowling Green State University and three summers at the University of Michigan Biology Station. His taxonomic breadth and passion for algae and teaching it to others was an enormous inspiration, and he made algae fun! The SEM images in this text were generated while at Bowling Green State University. The TEM images were generated while I worked as an Electron Microscopist in the Botany Department at The Ohio State University. A few of the images were taken while working for the United States Geological Survey.

Many of the light micrographs were generated from weekly class collections from the campus of Florida Gulf Coast University with over 400 acres of alkaline wetlands. I also benefited from colleagues who sent me live samples from across the country, particularly Oregon (Kurt Carpenter), Michigan (Kam Truhn), New Jersey (Ken Wagner), New York, Massachusetts, California, Washington, Utah, Idaho, Minnesota, Maryland, and more.

I also thank the reviewers of various taxonomic groups: Cyanobacteria—Jeffry Johansen; Euglenoids—Bożena Zakryś; Chrysophytes—Peter Siver; Diatoms—J. Partick Kociolek, Lane Allen, and Rex Lowe; Dinoflagellates—Matthew Parrow; Reds—Julie Hambrook Berkman and Morgan Vis-Chiasson; Greens—Thomas Frankovich.

I thank all the undergraduate and graduate students who helped me obtain live material. Seven undergraduates played a significant role in helping me assemble the images: Maura Cleary, Grace Demko, Enzo Ferraz, Veronica Salvatierra, Nyx Schuler, Allison Sanchez, and Rachel Pavel. This guide is also dedicated to the memory of Dr. Ann St. Amand, a fellow phycologist and workshop instructor for more than 30 years.

Contents

1 Cyanobacteria (Blue-Green Algae)

Introduction to the Cyanobacteria

Cyanobacteria are gram-negative bacteria that are photosynthetic, having chlorophyll *a*, that unites these prokaryotic organisms with the eukaryotic organisms, collectively called algae. As prokaryotes, the internal structure has no membrane-bound organelles, such as the chloroplasts, nucleus, and mitochondria. However, they do have structures including ribosomes, cyanophycean granules, polyhedral bodies, polyphosphate bodies, and some genera have gas vesicles. They also have membranes, such as the thylakoids embedded with chlorophyll molecules for light capture. Many of these structures are visible with light microscopy and aid in determining the genus of an organism. The internal cellular material is contained by a cytoplasmic membrane, which is then surrounded by a complex layered cell wall composed of mucopeptides, carbohydrates, amino acids, and fatty acids. Cyanobacteria also have extracellular polymeric substances that are often apparent and are composed of polysaccharides, proteins, lipids, and nucleic acids that are secreted by the cells and are used to determine a genus.

Cyanobacteria do not reproduce sexually; cells simply divide by a process called binary fission, with two identical daughter cells forming from the parent cell. There are other important characteristics that help distinguish cyanobacteria from the eukaryotic algae. For example, they never have flagellum, a feature common to several eukaryotic algae; however, some can "glide" when in contact with a surface. Cyanobacteria also have other light harvesting and photoprotective pigments, such as phycocyanin and phycoerythrin, along with various carotenoids, that impart color to cells ranging from blue-green to red, pink, purple, golden, brown, and black. Prior light conditions and nutritional status may influence the color of cells. When a cyanobacteria bloom deteriorates, the water-soluble phycocyanin leaks from cells as they die, imparting a brilliant turquoise surface sheen that is easily recognizable.

Morphologically, the simplest cyanobacteria are spherical, solitary cells that appear mostly round with the light microscope, surrounded by a thin layer of

B. H. Rosen, *Color Atlas of Freshwater Algae*,
https://doi.org/10.1007/978-3-031-93122-2_1

mucilage that is too transparent to see. Other solitary cyanobacteria are rod shaped (geometrically a capsule). Spherical and capsular organisms can live in a collective mucilaginous colony, which are visible and distinguish these into distinct genera. Cell shapes vary in the colonial cyanobacteria, and often, the number of cells in a colony is prolific and indeterminate, such as the bloom-former *Microcystis*. In addition of colonial cyanobacteria, there are filamentous forms where cells attach end to end, with short to long cells. One order, the Nostocales, form specialized cells for nitrogen fixation (heterocytes) and also can produce resting cells (akinetes). The most sophisticated forms are branched and multiseriate, with multiple cells in cross section.

Cyanobacteria

Some genera of cyanobacteria form only single cells that don't have enough visible characteristics to identify, and genetic information is needed to place them in the correct genus. As an example, morphologically, this organism fits into *Cyanobacterium cedrorum*; however, the taxonomic position of this species is unresolved, and it may eventually be assigned to a different order, family, or genus.

The morphology represented in the images in the Atlas is used to classify these organisms as a starting point. Each taxon is reviewed for its current taxonomic position with the knowledge that refinements and reassignments are leading to new names and lineages revealed through genetic analyses. Therefore, the names assigned by the author are valid at the time of publication and are subject to change (Figs. 1.1 to 1.111).

Class: Cyanophyceae
Order: Chroococcales
Family: Geminocystaceae

Cyanobacteria

Cyanobacterium

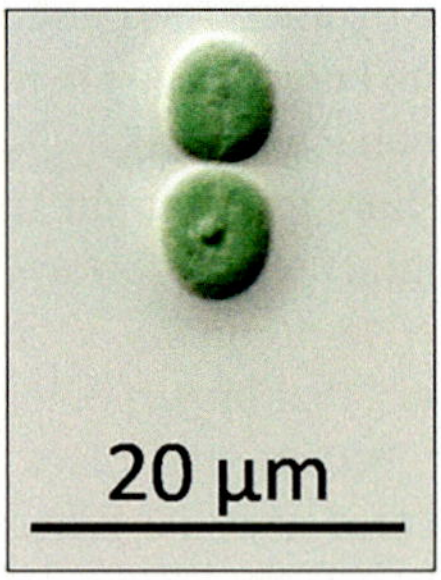

Cyanobacterium cedrorum Metaphyton and periphyton

Fig. 1.1 *Cyanobacterium*

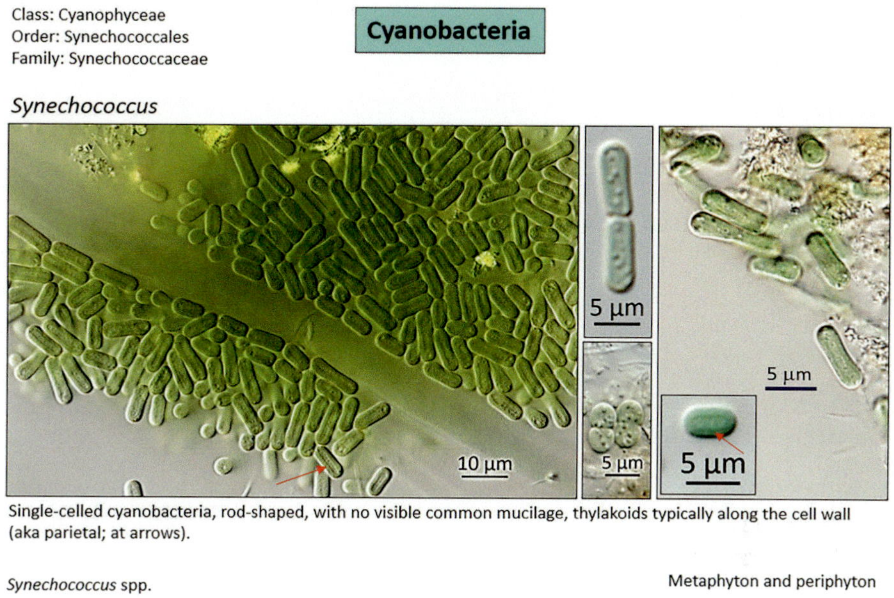

Fig. 1.2 *Synechococcus*

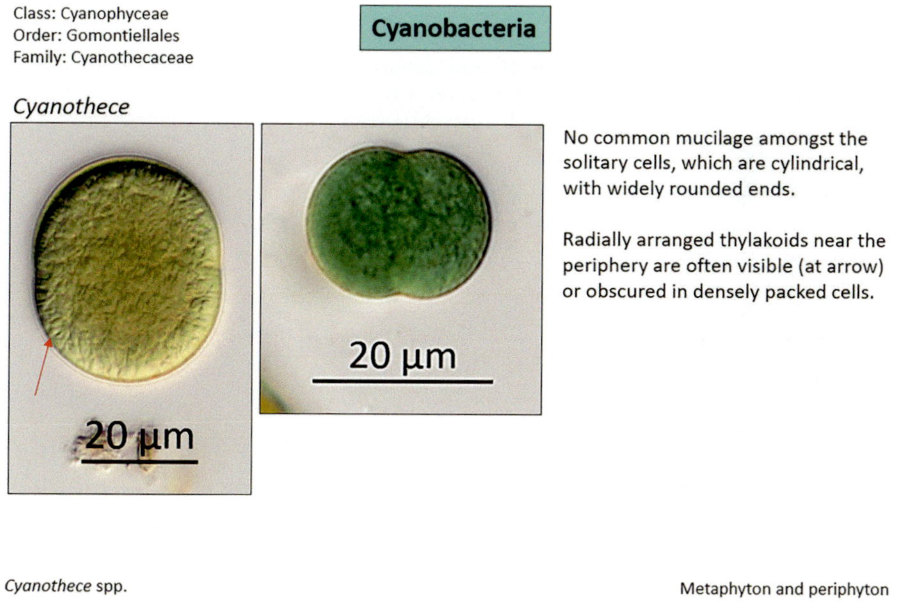

Fig. 1.3 *Cyanothece*

Class: Cyanophyceae
Order: Chroococcales
Family: Microcystaceae

Cyanobacteria

Synechocystis

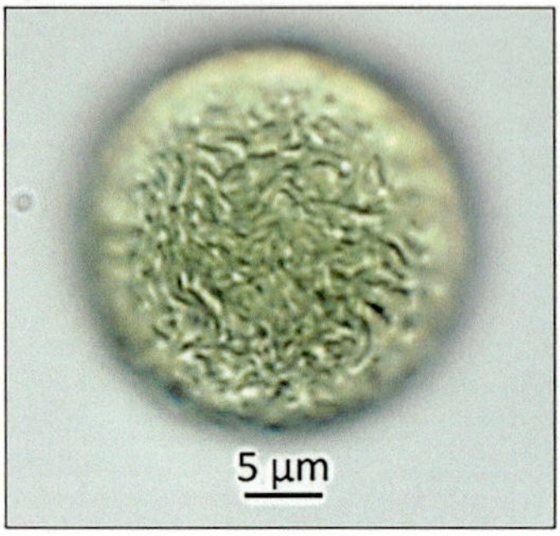

Cells spherical and solitary, without mucilage.

Synechocystis sallensis

Metaphyton and periphyton

Fig. 1.4 *Synechocystis*

Class: Cyanophyceae
Order: Chroococcales
Family: Microcystaceae

Cyanobacteria

Geminocystis

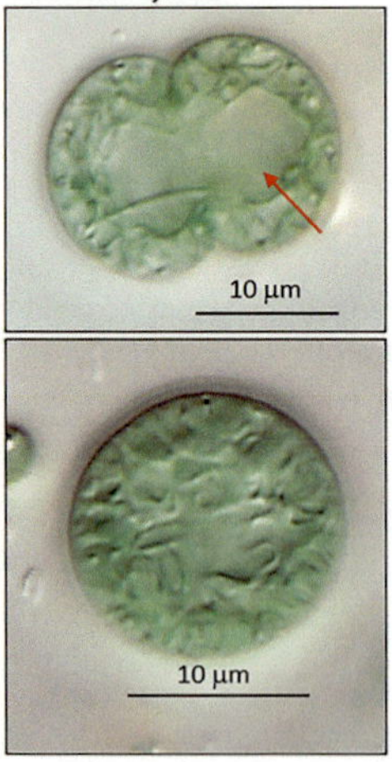

Cells spherical, often with kerotimization, leaving a widened area that has the appearance of a vacuole (arrow)

Geminocystis sp.

Metaphyton and periphyton

Fig. 1.5 *Geminocystis*

Class: Cyanophyceae
Order: Chroococcales
Family: Microcystaceae

Cyanobacteria

Aphanocapsa

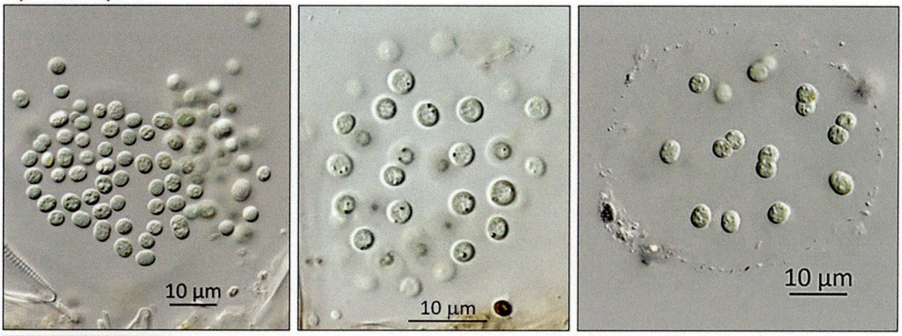

Spherical cells held together by a gelatinous matrix that may not be obviously visible; cell in the process of dividing are elongated before fission.

Aphanocapsa spp.

Plankton, metaphyton and periphyton

Fig. 1.6 *Aphanocapsa*

Class: Cyanophyceae
Order: Chroococcales
Family: Microcystaceae

Cyanobacteria

Aphanocapsa

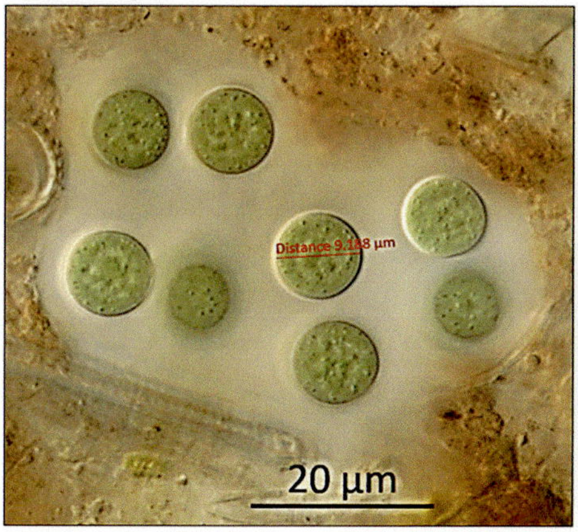

Aphanocapsa sesciacensis

imbedded in the periphytic community

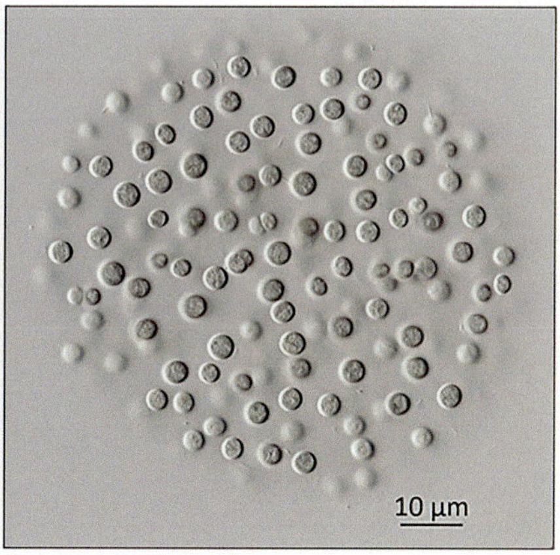

Aphanocapsa planctonica

a common planktonic form

Plankton, metaphyton and periphyton

Fig. 1.7 *Aphanocapsa*

Class: Cyanophyceae
Order: Chroococcales
Family: Microcystaceae

Cyanobacteria

Aphanothece

Oval to rod-shaped cells held together by a gelatinous matrix that may not be obviously visible.

Aphanothece spp. Plankton, metaphyton and periphyton

Fig. 1.8 *Aphanothece*

Class: Cyanophyceae
Order: Nodosilineales
Family: Cymatolegaceae

Cyanobacteria

Lemmermanniella

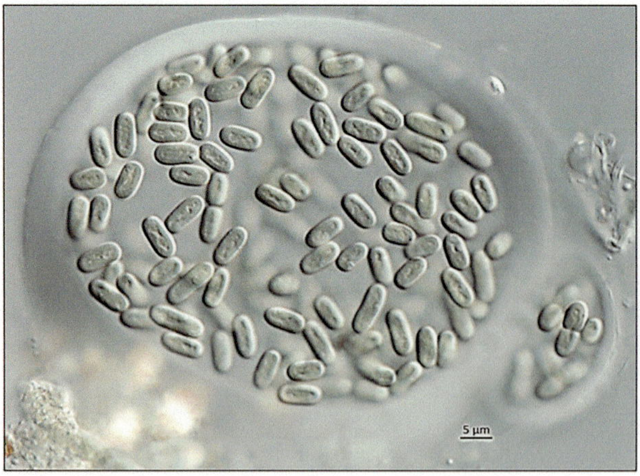

Thick mucilaginous spherical or oval colony with rod-shaped cells forming a single layer.

Lemmermanniella uliginosa Tropical and subtropic alkaline wetlands

Fig. 1.9 *Lemmermanniella*

Class: Cyanophyceae
Order: Chroococcales
Family: Microcystaceae

Cyanobacteria

Gloeothece

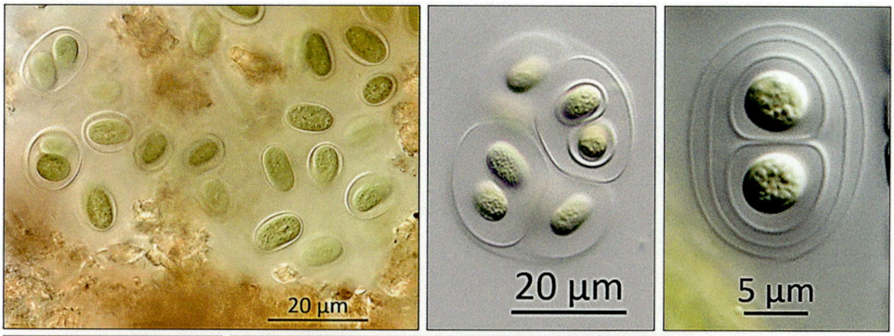

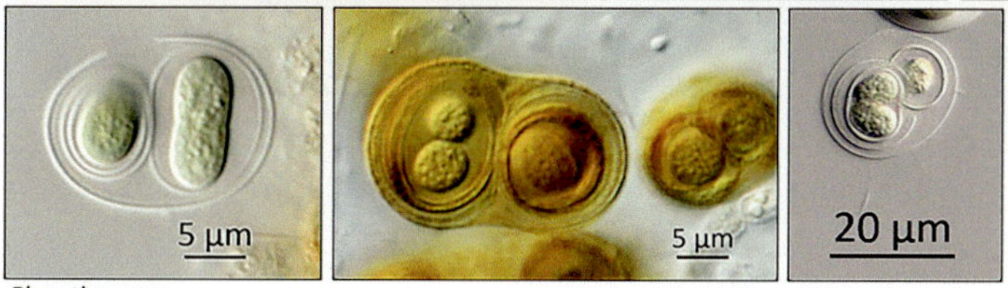

Oval to rod-shaped cells with individual gelatinous envelopes and collective mucilage.

Gloeothece spp.

Aerophilic, epilithic, plankton, metaphyton and periphyton

Fig. 1.10 *Gloeothece*

Class: Cyanophyceae
Order: Chroococcales
Family: Microcystaceae

Cyanobacteria

Eucapsis

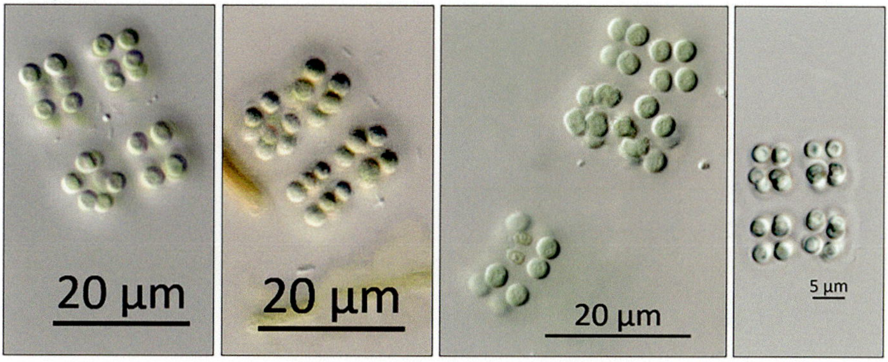

Cuboidal colonies held together by a common mucilaginous sheath, with 4 cells per face.

Eucapsis spp.

Plankton, metaphyton and periphyton

Fig. 1.11 *Eucapsis*

Class: Cyanophyceae
Order: Chroococcales
Family: Microcystaceae

Cyanobacteria

Coelosphaerium

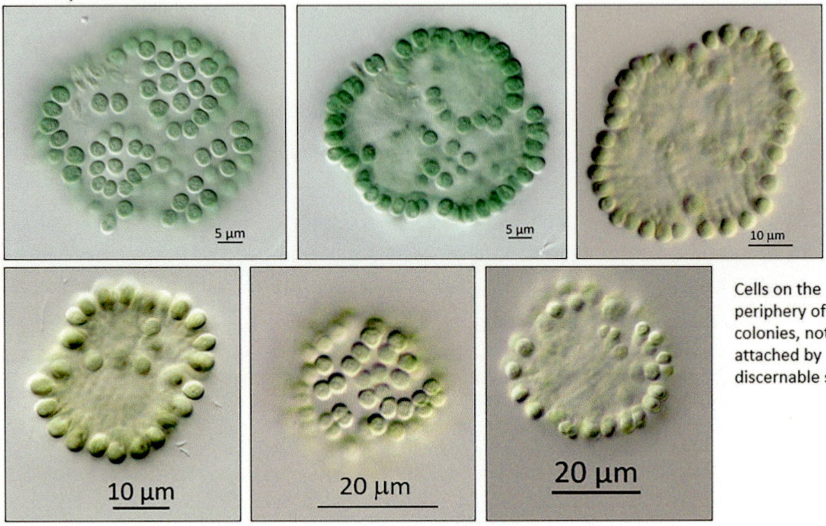

Cells on the periphery of colonies, not attached by discernable stalks.

Coelosphaerium kuetzingianum

Plankton

Fig. 1.12 *Coelosphaerium*

Class: Cyanophyceae
Order: Chroococcales
Family: Microcystaceae

Cyanobacteria

Snowella

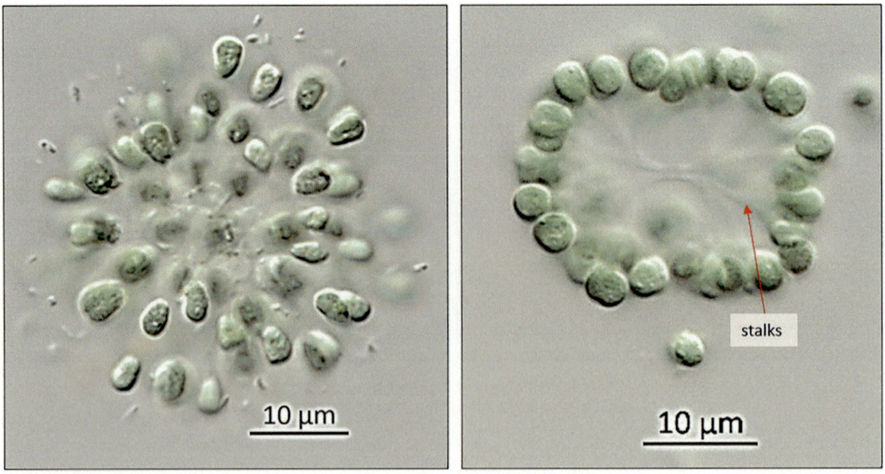

Snowella lacustris

Snowella litoralis

Plankton

Fig. 1.13 *Snowella*

Class: Cyanophyceae
Order: Chroococcales
Family: Microcystaceae

Cyanobacteria

Merismopedia

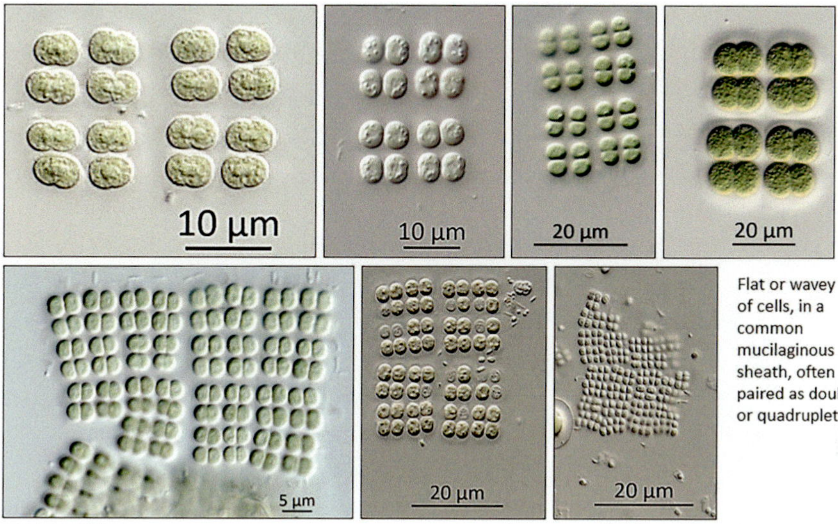

Flat or wavey plates of cells, in a common mucilaginous sheath, often paired as doublets or quadruplets.

Merismopedia spp.

Plankton, metaphyton and periphyton

Fig. 1.14 *Merismopedia*

Class: Cyanophyceae
Order: Chroococcales
Family: Chroococcaceae

Cyanobacteria

Chroococcus

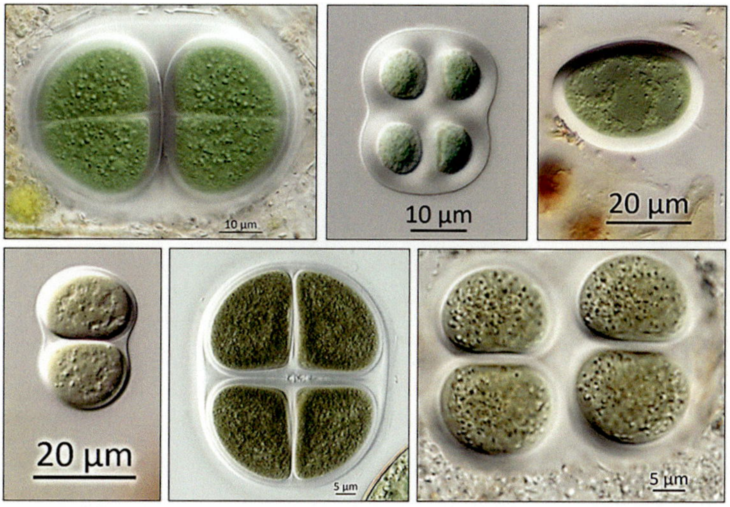

Hemispherical cells with individual gelatinous or collective mucilage. Exhibit a wide range of color variations and habitats.

Chroococcus spp.

Plankton, metaphyton and periphyton

Fig. 1.15 *Chroococcus*

Class: Cyanophyceae
Order: Chroococcales
Family: Chroococcaceae

Cyanobacteria

Cyanokybus

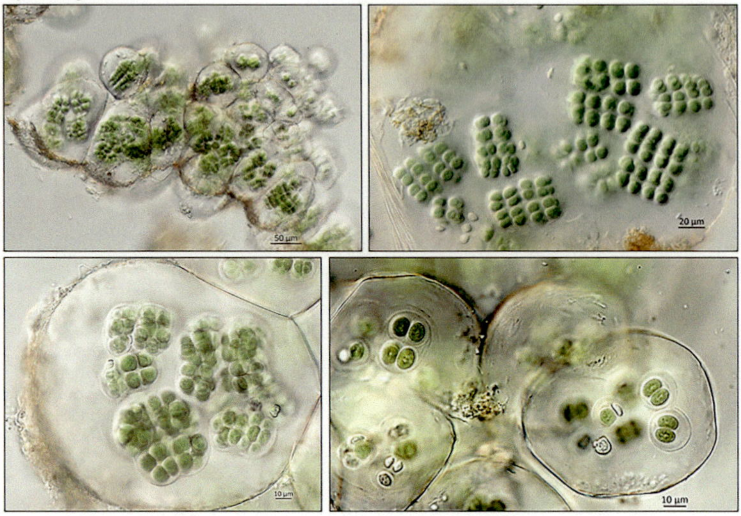

Hemispherical cells, often in pairs with gelatinous, arranged in rows surrounded by collective mucilage and clustered.

Cyanokybus venezuelae

Temporary wetland in the subtropics

Fig. 1.16 *Cyanokybus*

Class: Cyanophyceae
Order: Chroococcales
Family: Chroococcaceae

Cyanobacteria

Cyanosarcina

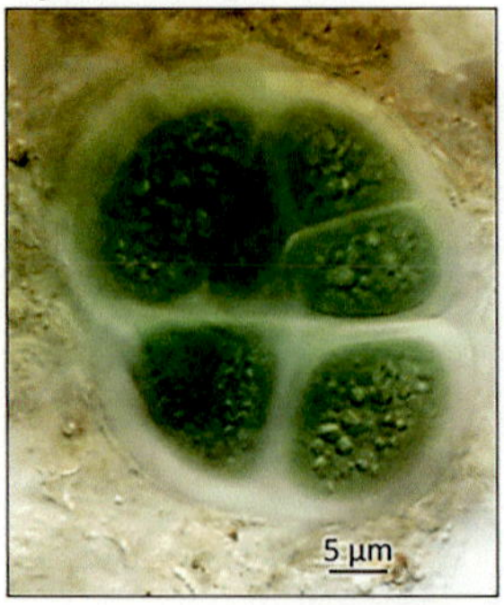

Cyanosarcina chroococcoides

Pseudocapsa

Pseudocapsa venkataramanii

Periphyton

Fig. 1.17 *Cyanosarcina* and *Pseudocapsa*

Class: Cyanophyceae
Order: Chroococcales
Family: Cyanothrichaceae

Cyanobacteria

Limnococcus

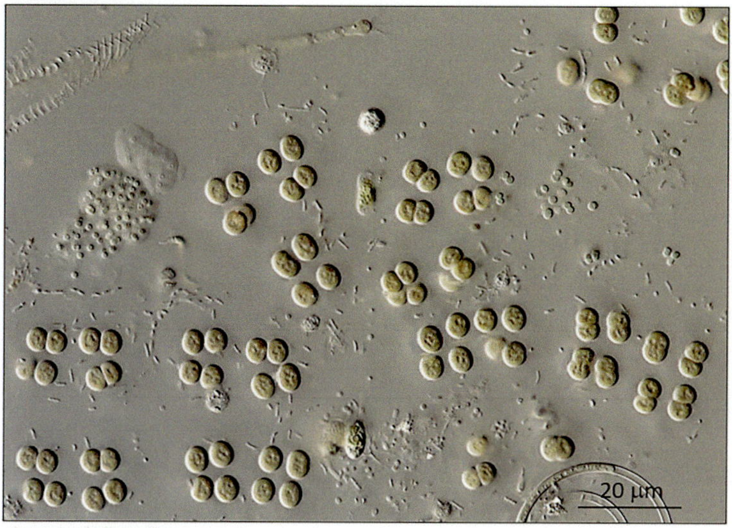

Similar in appearance *Chroococcus*, however cells are spherical when full formed.

Limnococcus limneticus

Plankton

Fig. 1.18 *Limnococcus*

Class: Cyanophyceae
Order: Chroococcales
Family: Cyanothrichaceae

Cyanobacteria

Johannesbaptistia

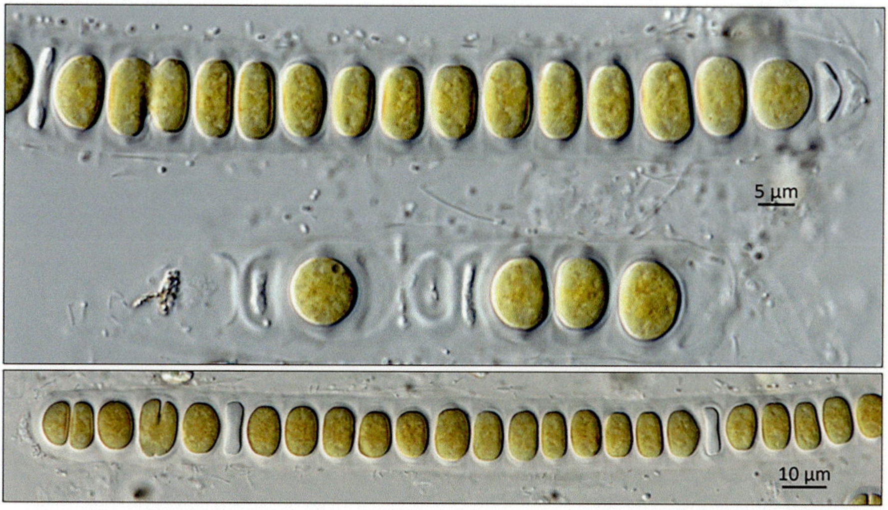

Johannesbaptistia pellucida

Periphyton

Fig. 1.19 *Johannesbaptistia*

Class: Cyanophyceae
Order: Chroococcales
Family: Gomphosphaeriaceae

Cyanobacteria

Gomphosphaeria

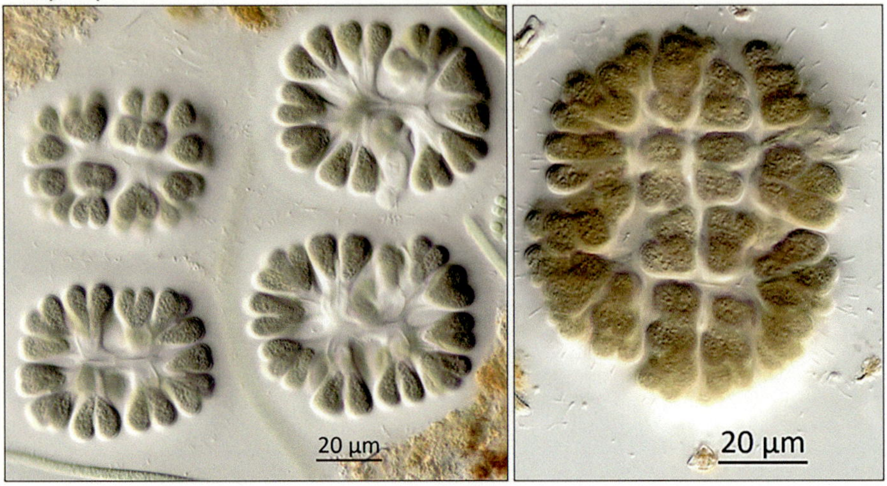

Gomphosphaeria aponina

Metaphyton and often floating

Fig. 1.20 *Gomphosphaeria*

Class: Cyanophyceae
Order: Chroococcales
Family: Gomphosphaeriaceae

Cyanobacteria

Gomphosphaeria

Colonies are spherical or oval or irregular. Cells attached to thick stalks that are branched and terminate in cells, often heart-shaped (cordiform).

Gomphosphaeria natans

Metaphyton and often floating

Fig. 1.21 *Gomphosphaeria*

Class: Cyanophyceae
Order: Chroococcales
Family: Pleurocapsaceae

Cyanobacteria

Chroococcopsis

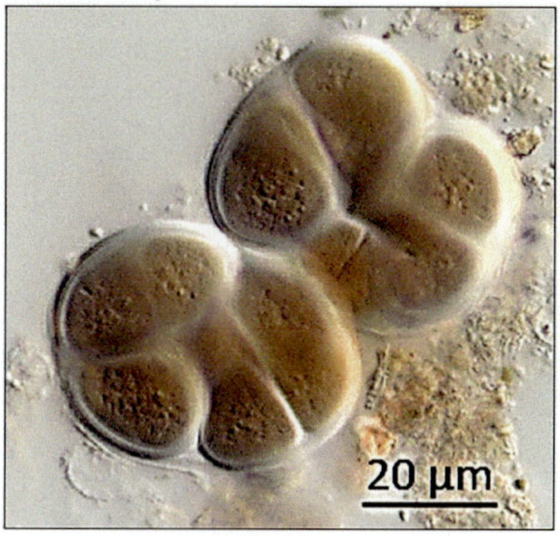

Irregular shaped cells with rounded ends, in a colony surrounded by a clear envelop. Reddish color due to phycoerythrin and/or carotenoids.

Chroococcopsis sp. Periphyton

Fig. 1.22 *Chroococcopsis*

Class: Cyanophyceae
Order: Chroococcales
Family: Pleurocapsaceae

Cyanobacteria

Onkonema (cf)=difficulty to form a positive identification)

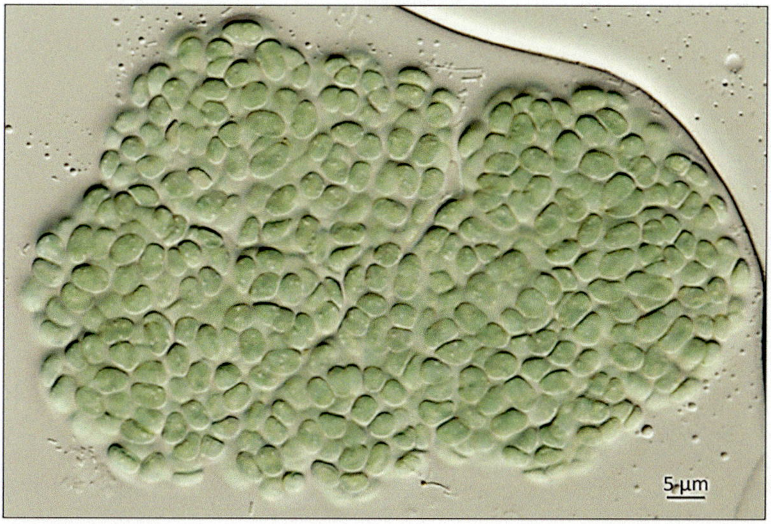

Onkonema paludosum Calcareous tropical wetlands-periphyton

Fig. 1.23 *Onkonema* (cf = difficulty to form a positive identification)

Class: Cyanophyceae
Order: Chroococcales
Family: Pleurocapsaceae

Cyanobacteria

Cyanocystis

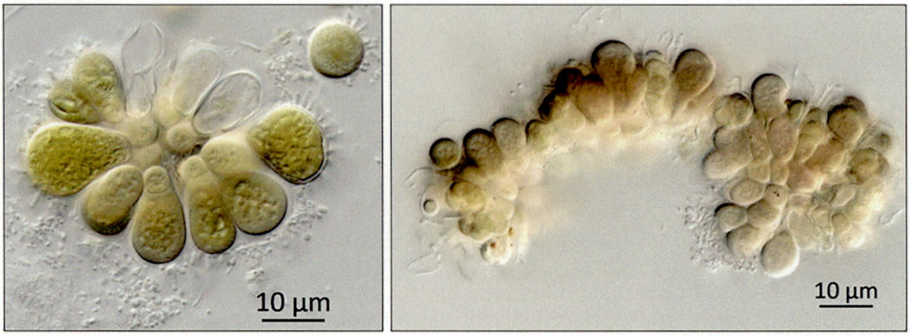

Pear-shaped cells, in a semicircle.

Cyanocystis mexicana

Calcareous tropical wetlands-periphyton

Fig. 1.24 *Cyanocystis*

Class: Cyanophyceae
Order: Chroococcidiopsidales
Family: Gloeocapsaceae

Cyanobacteria

Chlorogloea

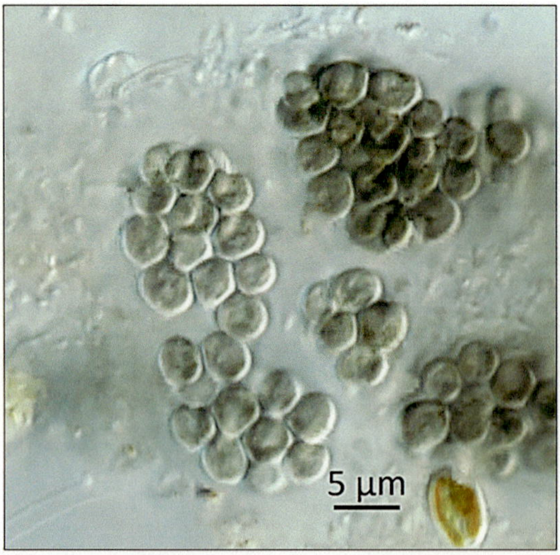

Chlorogloea purpurea

Calcareous tropical wetlands-periphyton

Fig. 1.25 *Chlorogloea*

Class: Cyanophyceae
Order: Pleurocapsales
Family: Hyellaceae

Cyanobacteria

Chamaecalyx

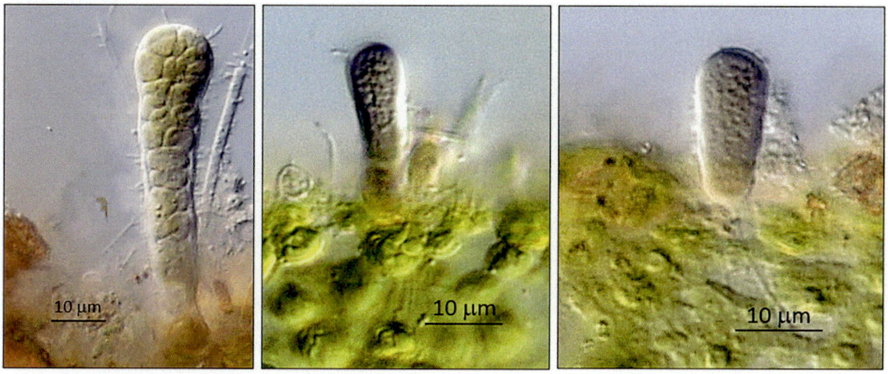

Individual cells attached to other algae and plants by a pad. Cells are club-shape and reproduction is visible in the image on the left, with formation of daughter cells (exocytes).

Chamaecalyx swirenkoi

Epiphytic on subtropical filamentous green algae

Fig. 1.26 *Chamaecalyx*

Class: Cyanophyceae
Order: Gomontiellales
Family: Chamaesiphonaceae

Cyanobacteria

Stichosiphon

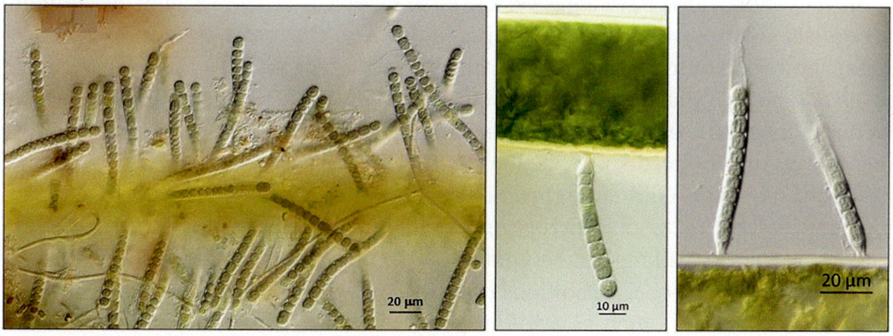

Short erect trichomes, attached by a basal cell differentiated for attachment.

Stichosiphon willei

Epiphytic

Fig. 1.27 *Stichosiphon*

Class: Cyanophyceae
Order: Nodosilineales
Family: Cymatolegaceae

Cyanobacteria

Romeria

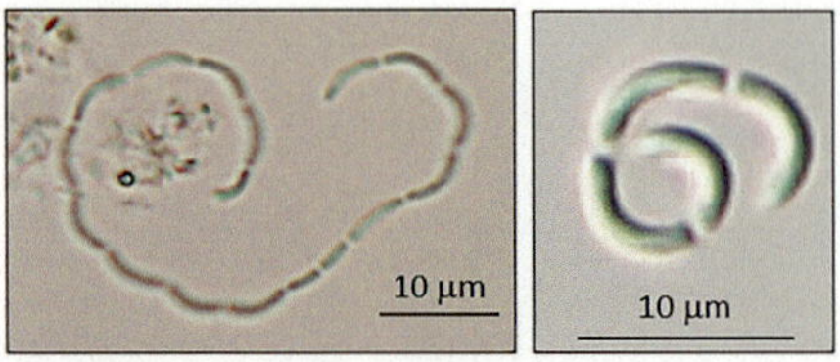

Romeria spp.

Relatively rare planktonic pseudofilament with disjointed curved cells. No aerotopes.

Wolskyella

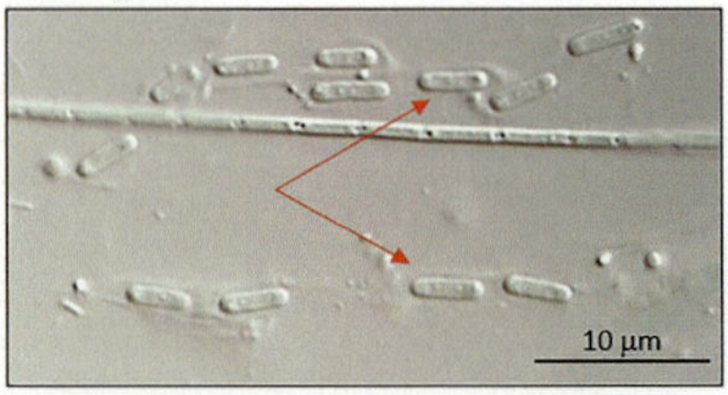

Wolskyella floridana at arrows. Image originally published in https://doi.org/10.3133/ofr20161114

A pseudofilamentous genus has a colorless mucilaginous envelop with cylindrical cells, distant from one another. Collected from the USA (Florida Everglades) periphyton.

Metaphyton, benthic

Fig. 1.28 *Romeria* and *Wolskyella*

Harmful Algal Blooms

Harmful algal blooms (HABs), a term with a variety of definitions and meanings, with no exact agreement what constitutes a HAB. There are several genera of cyanobacteria that are planktonic and form visible floating masses that are the easiest to recognize, and cyanobacterial blooms are termed "cyanoHABs" (cHABs). Most of these cyanobacteria have a gas vesicle, also known as aerotopes, that allows them to regulate their buoyancy, moving up and down in the water column to optimize light capture and nutrient harvesting from deeper waters, and this gives these genera an advantage over many other planktonic algae. However, when a essential element, such as phosphorus or nitrogen, or other environmental condition becomes limiting and causes metabolic disruption, the floating masses can become trapped at the surface be overexposed to sunlight and oxidize critical pigments, creating a bleached appearance. This metabolic disruption can also lead to the collapse of the bloom, with some genera sinking to the benthos, forming resting structures or quiescent forms that resurface when conditions improve. Not all cyanobacteria with aerotopes form surface blooms, such as *Raphidiopsis,* which blooms in mid-water column. Another category of cyanobacteria that form HABs live attached to hard surfaces in streams and rivers and other benthic habitats in waterbodies, forming mats that can break free and float as well. Eukaryotic algae that form mats or clog waterways are often termed "nuisance algae" blooms, mostly because these do not produce cyanotoxins.

The majority harm from cHAB is caused by low dissolved oxygen (DO) events. In a thriving bloom, photosynthesis oxygenates the surrounding water during the day, which is offset to a small extent by cellular respiration that consumes oxygen. At night, the cyanobacteria in the bloom utilize stored sugars during respiration which consumes oxygen from the water. When a bloom covers the surface of a waterbody, a low DO event can cover a large enough area that fish and invertebrates are killed because they cannot move to oxygenic waters. In addition, when a cHAB stops thriving and enters the collapse phase, the cellular contents becomes readily available to feed these heterotrophic bacteria, causing more extensive fish kills.

Several cHABs can make cyanotoxins, naturally occurring compounds that have important cellular functions, such as antioxidants, nitrogen storage, and other metabolic benefits to be discovered. For animals, especially mammals, these cyanotoxins have detrimental effects if ingested. They are neurotoxic, hepatoxic, and cytotoxic and require genes (activated) for cyanotoxin synthesis. It is not accurate to assume that the dominant organism in a bloom is the cyanotoxin producer, and the identification of the members of a bloom is not sufficient to identify a toxin producer. Undoubtedly, more cyanobacteria will be identified that synthesize toxins that are not identified herein.

Class: Cyanophyceae
Order: Chroococcales
Family: Microcystaceae

Cyanobacteria
cHAB

Microcystis

Microcystis spp. are a colonial cyanobacteria.

This species, visible without magnification, is identified by open spaces within the clusters of cells (clathrate).

Several strains of this species have the ability to produce one or more of the 300 variants of microcystin, some with known liver toxicity.

Microcystis aeruginosa microcystins Plankton

Fig. 1.29 *Microcystis*

Class: Cyanophyceae
Order: Chroococcales
Family: Microcystaceae

Cyanobacteria
cHAB

Microcystis

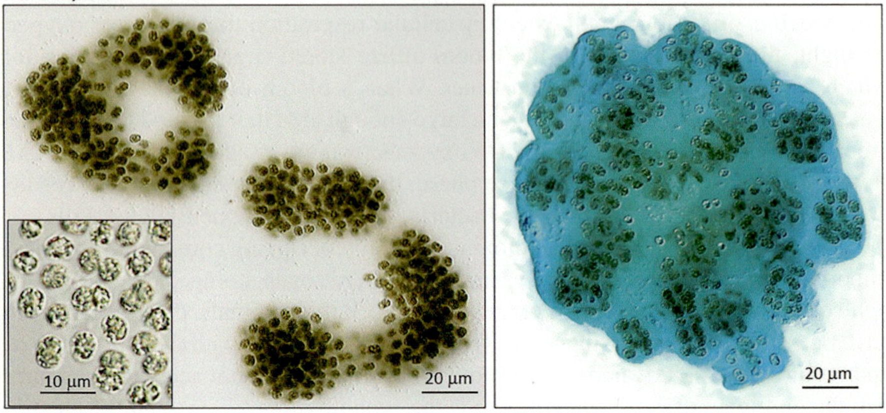

Cells aerotopes have a "grainy" appearance (insert). Colonies are held together by a common mucilage, not readily apparent but highly visible when stained with Alcian Blue (on right).

Microcystis aeruginosa microcystins Plankton

Fig. 1.30 *Microcystis*

Class: Cyanophyceae
Order: Chroococcales
Family: Microcystaceae

Cyanobacteria
cHAB

Microcystis

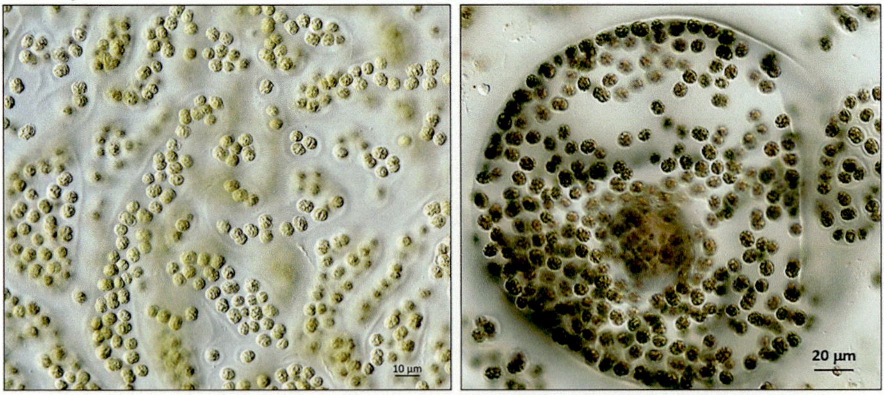

This species of *Microcystis* secretes a firm mucilage layer that is easy to discern under the microscope. The cells are are not embedded in the mucilage and can move freely inside the colony.

Microcystis wesenbergii microcystins Plankton

Fig. 1.31 *Microcystis*

Class: Cyanophyceae
Order: Chroococcales
Family: Microcystaceae

Cyanobacteria
cHAB

Microcystis

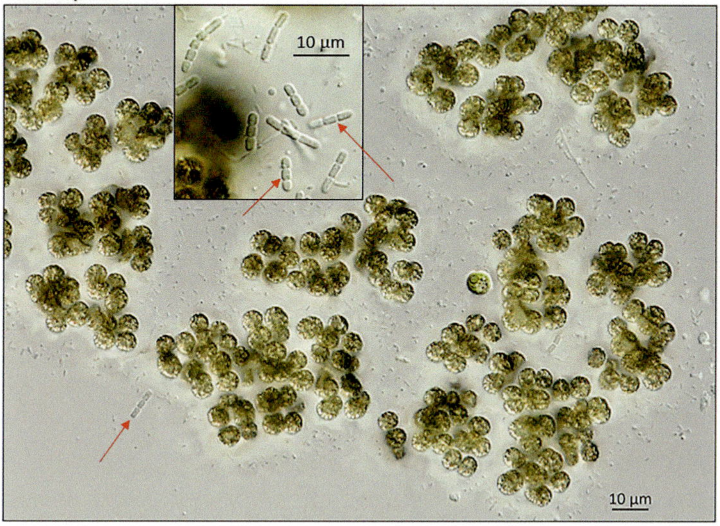

This species of *Microcystis* has cells that arrange in packets that often look like cubes.

An inhabitant of *Microcystis* spp. mucilage (arrows/insert) *Pseudanabaena mucicola*

Microcystis viridis microcystins Plankton

Fig. 1.32 *Microcystis*

Class: Cyanophyceae
Order: Chroococcales
Family: Microcystaceae

Cyanobacteria
cHAB

Microcystis

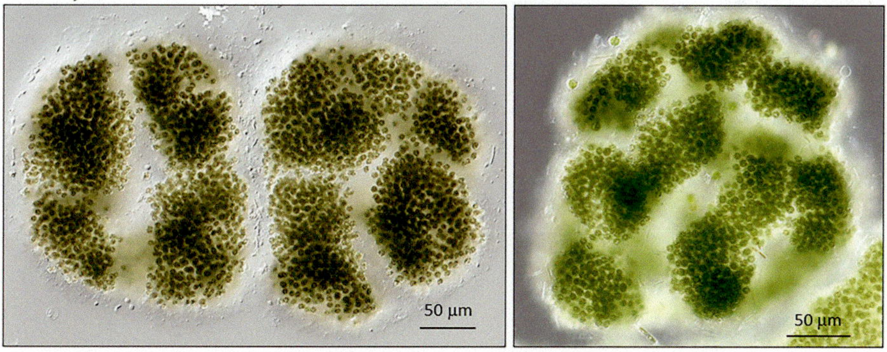

This species of *Microcystis* forms colonies with tightly packed cells, with clusters of cells in older colonies.

Microcystis novacekii microcystins Plankton

Fig. 1.33 *Microcystis*

Class: Cyanophyceae
Order: Chroococcales
Family: Microcystaceae

Cyanobacteria
cHAB

Microcystis

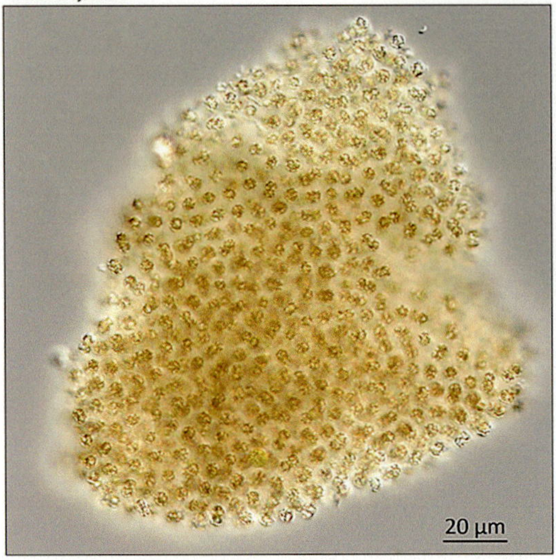

This species of *Microcystis* has densely arranged cells.

Microcystis flos-aquae microcystins Plankton

Fig. 1.34 *Microcystis*

Class: Cyanophyceae
Order: Synechococcales
Family: Coelosphaeriaceae

Cyanobacteria
cHAB

Woronichinia

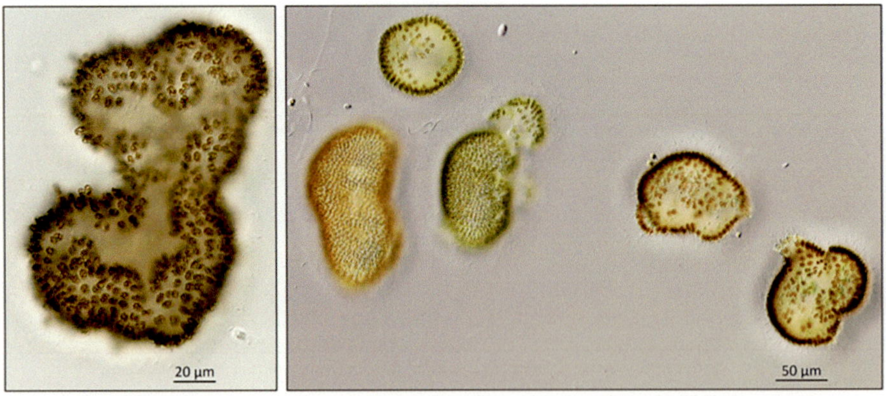

Cells arranged as a single layer surround a central mass of mucilage. Colonies spherical to highly irregular. A common bloom-former.

Woronichinia naegeliana cyanopeptides Plankton

Fig. 1.35 *Woronichinia*

Class: Cyanophyceae
Order: Synechococcales
Family: Coelosphaeriaceae

Cyanobacteria
cHAB

Woronichinia

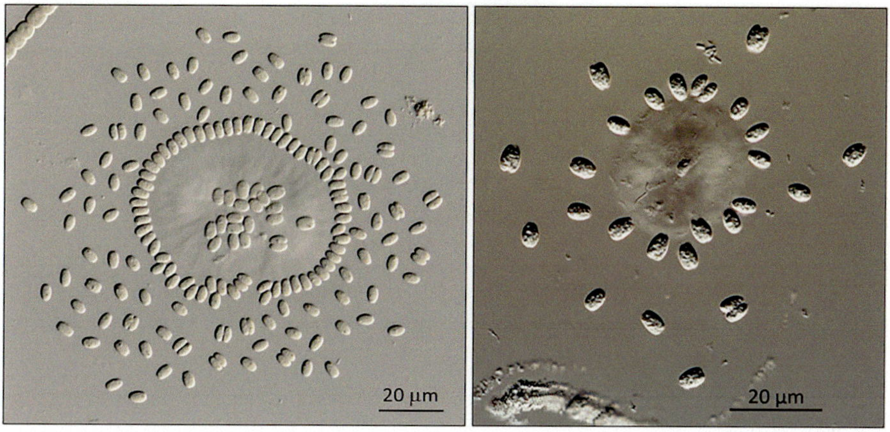

Cells dispersing from the central mucilage.

Woronichinia naegeliana cyanopeptides Plankton

Fig. 1.36 *Woronichinia*

Class: Cyanophyceae
Order: Gomontiellales
Family: Gomontiellaceae

Cyanobacteria

Borzia

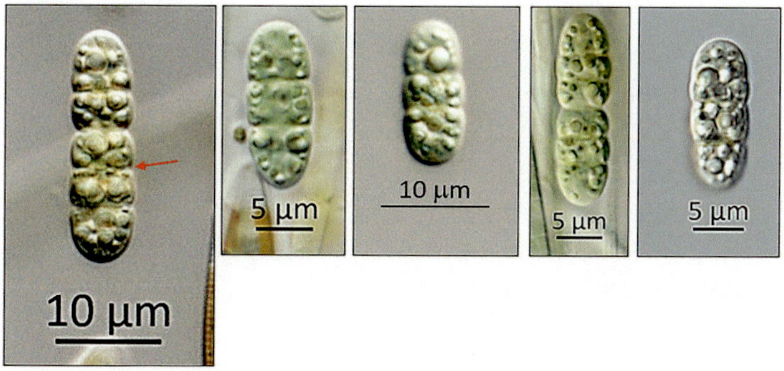

The linear arrangement of cells (a trichome) in this genus, has a few cells constricted at the cross walls (arrow) and numerous inclusions that appear circular or amorphous.

Borzia trilocularis Periphyton

Fig. 1.37 *Borzia*

Class: Cyanophyceae
Order: Gomontiellales
Family: Gomontiellaceae

Cyanobacteria

Komvophoron

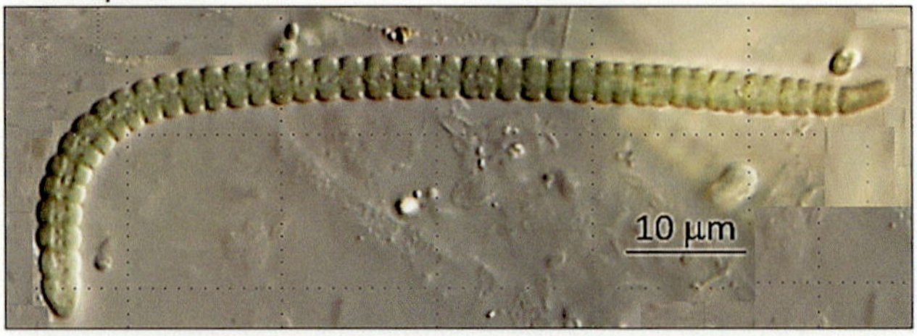

Komvophoron rostratum

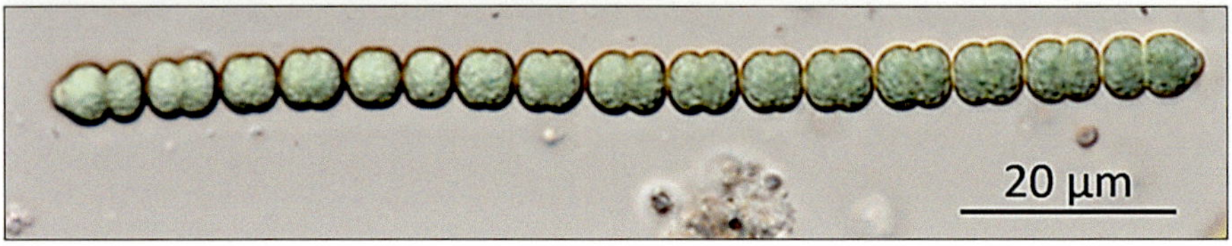

Komvophoron schmidlei

Plankton and periphyton

Fig. 1.38 *Komvophoron*

Class: Cyanophyceae
Order: Spirulinales
Family: Spirulinaceae

Cyanobacteria

Spirulina

Spirulina spp.

Plankton and periphyton

Fig. 1.39 *Spirulina*

Class: Cyanophyceae
Order: Spirulinales
Family: Spirulinaceae

Cyanobacteria

Spirulina

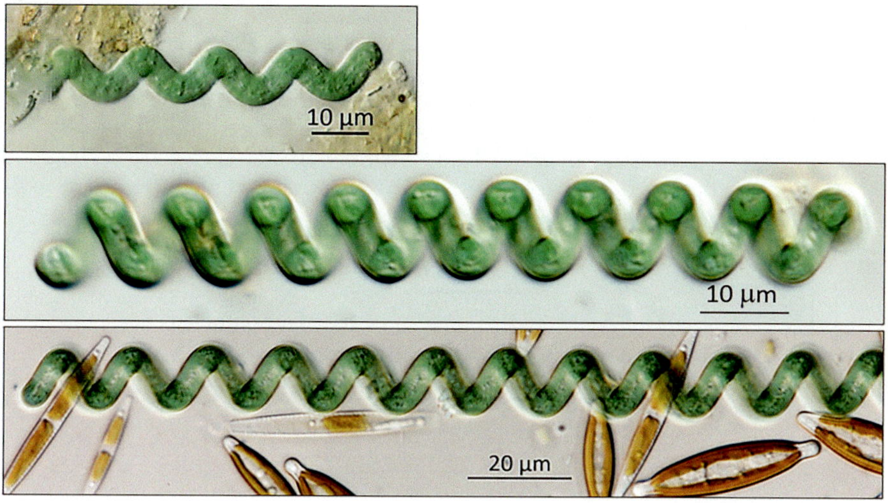

Spirulina meneghiniana

Plankton and periphyton

Fig. 1.40 *Spirulina*

Cyanobacteria Order Nostocales

The remainder of this section covers the most complex group of filamentous cyanobacteria based on morphology, specialized cells, and metabolic function. Approximately, 1800 species are currently recognized in the Order Nostocales, including many that synthesize compounds that are considered cyanotoxins. Some members exist as single filaments, straight, coiled, sometimes tapered, and with various terminal cell morphology, while others group together and form bundles or colonies.

Members of this order have evolved to capture atmospheric nitrogen (N_2), giving this group an ecological advantage over most other cyanobacteria (with some exceptions) and eukaryotic algae. A sophisticated complex of enzymes, the nitrogenases, function inside a specialized cell, the heterocyte, which has undergone physical and metabolic changes to accommodate these enzymes. Nitrogenases are inhibited by oxygen and as the heterocyte from a vegetative cell in a filament, photosystem II, the oxygenic portion of photosynthesis, is shut down, while keeping energy generation portion of photosynthesis because of the high energy demand needed during fixation. Heterocytes typically have and alter pigment composition that is visible with the light microscope. In addition, the thickened outer wall of the heterocyte is visible and impregnated with lipid to prevent passive diffusion of oxygen from

Class: Cyanophyceae
Order: Spirulinales
Family: Spirulinaceae

Cyanobacteria

Glaucospira

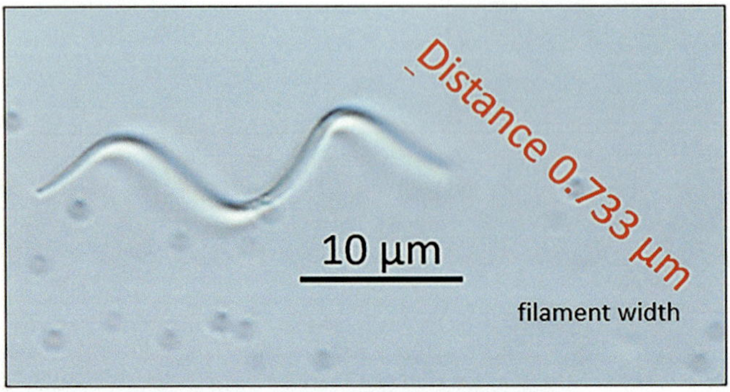

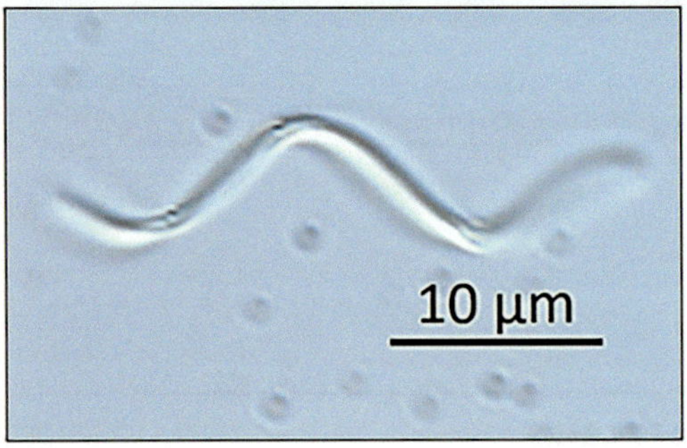

Glaucospira laxissima

Fig. 1.41 *Glaucospira*

entering. Heterocyte shape, size, and position in the trichome are some of the morphological attributes that help identify organisms.

Another specialized cell in most of the Nostocales is the akinete, which differentiates from a vegetative cell and acts as a resting stage that survives when the environmental conditions are suboptimal. Akinetes have protective outer layers and important storage molecules internally and can germinate after the environmental conditions are restored. Akinete size, shape, and position relative to the heterocyte are essential for the proper identification of many of the Nostocales.

Some members of the Nostocales also can form "branches," when a trichome develops in a different plane. False branching often occurs after a heterocyte, or as true branching, typically at right angles to the axis of the trichome. The most complex Nostocales are multiseriate, with cells dividing laterally along the length of the main trichome axis (Figs. 1.66 to 1.111).

Class: Cyanophyceae
Order: Oculatellales
Family: Oculatellaceae

Cyanobacteria

Elainella

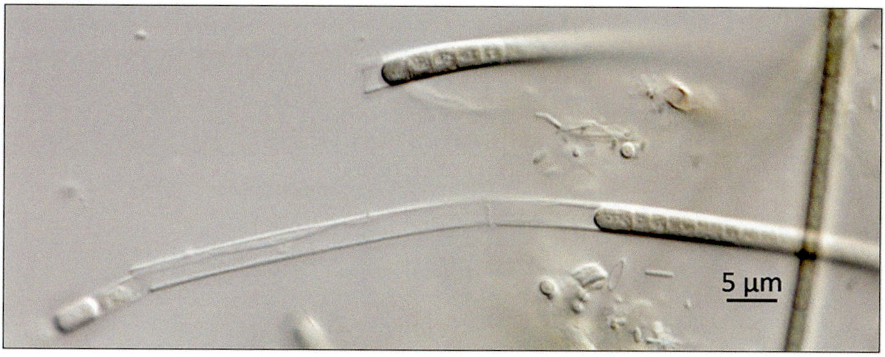

Elainella saxicola

Periphyton

Fig. 1.42 *Elainella*

Class: Cyanophyceae
Order: Leptolyngbyales
Family: Leptolyngbyaceae

Cyanobacteria

Leptolyngbya

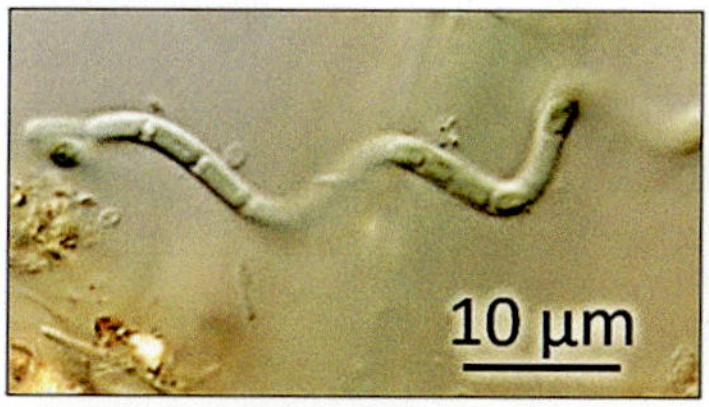

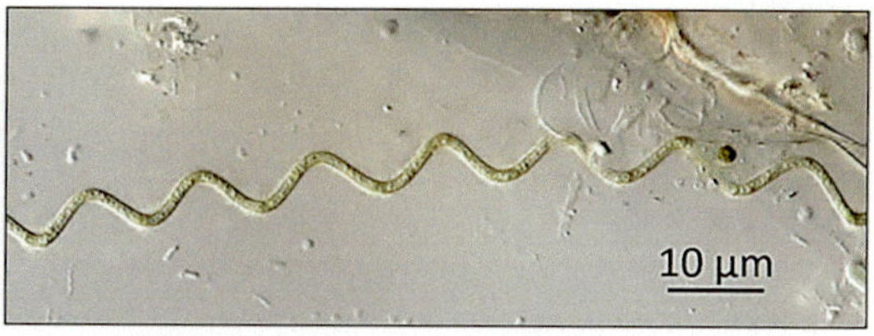

Leptolyngbya lagerheimii

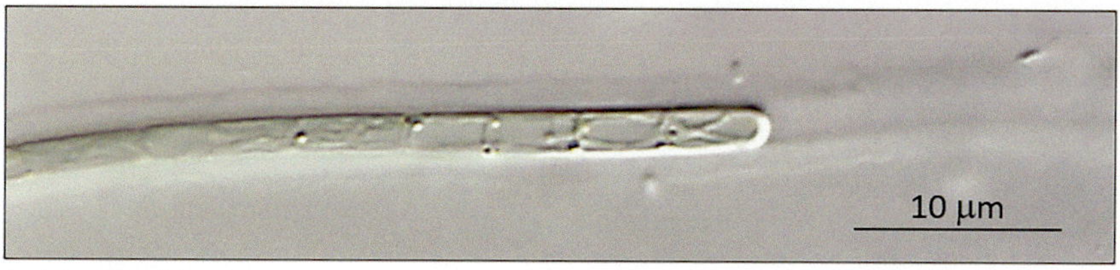

Leptolyngbya eliskae

Periphyton

Fig. 1.43 *Leptolyngbya*

Class: Cyanophyceae
Order: Leptolyngbyales
Family: Schizotrichaceae

Cyanobacteria

Geitleribacteron

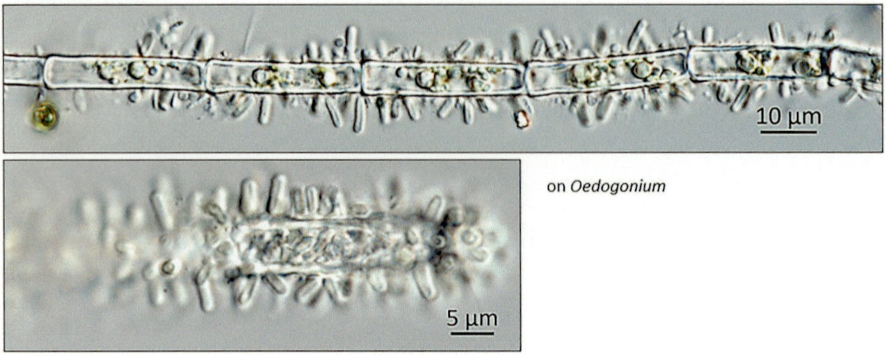

on *Oedogonium*

Geitleribactron periphyticum

epiphyte

Fig. 1.44 *Geutleribacteron*

Class: Cyanophyceae
Order: Leptolyngbyales
Family: Trichocoleusaceae

Cyanobacteria

Schizothrix

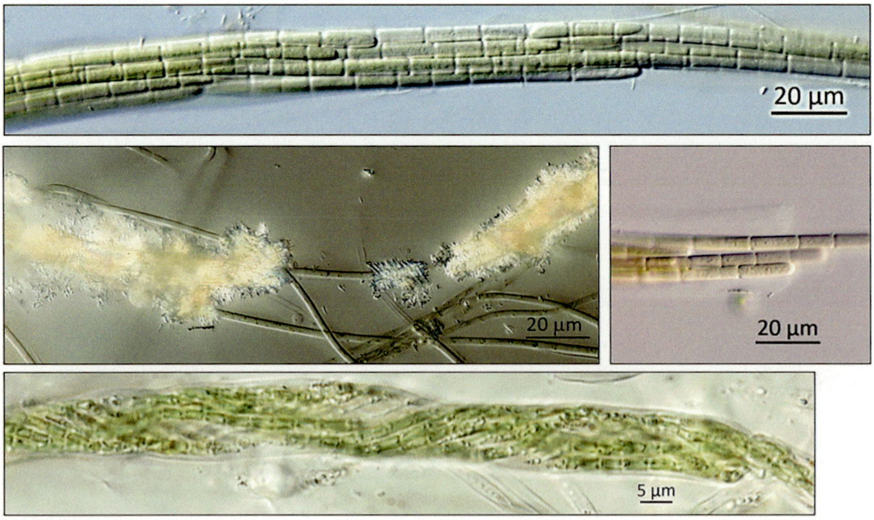

Schizothrix spp.

Periphyton

Fig. 1.45 *Schizothrix*

Class: Cyanophyceae
Order: Pseudanabaenales
Family:
Pseudanabaenaceae

Cyanobacteria
cHAB

Pseudanabaena

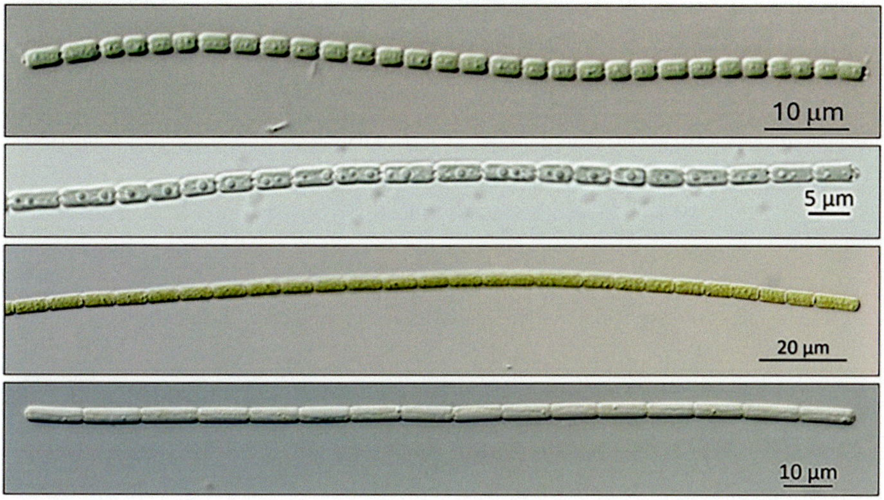

Pseudanabaena spp. anatoxin Plankton and periphyton

Fig. 1.46 *Pseudanabaena*

Class: Cyanophyceae
Order: Pseudanabaenales
Family: Pseudanabaenaceae

Cyanobacteria
cHAB

Limnothrix

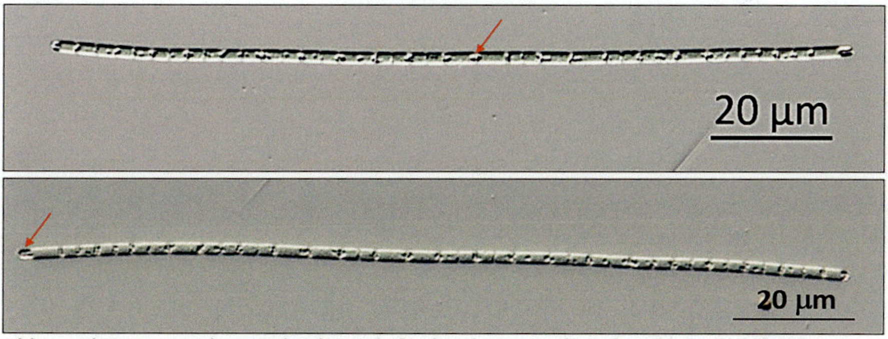

Trichomes show aerotopes (at arrows) at the terminal ends and at cross wells. Toxin activity has been found; however, the exact form of toxin has not been fully documented.

Limnothrix redekei (Goor) Meffert 1988: 274 saxitoxins Plankton and periphyton

Fig. 1.47 *Limnothrix*

Class: Cyanophyceae
Order: Oscillatoriales
Family: Microcoleaceae

Cyanobacteria
cHAB

Planktothrix

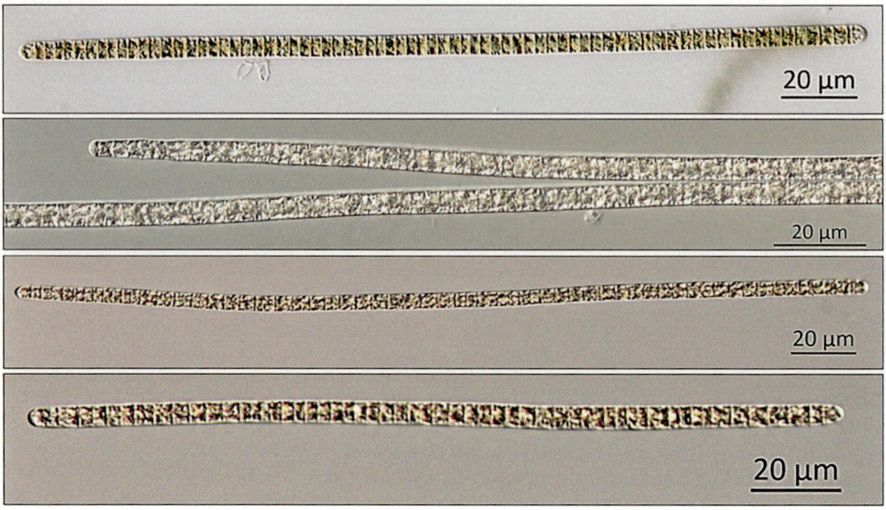

Planktothrix spp. Microcystins Plankton

Fig. 1.48 *Planktothrix*

Class: Cyanophyceae
Order: Oscillatoriales
Family: Microcoleaceae

Cyanobacteria

Trichodesmium

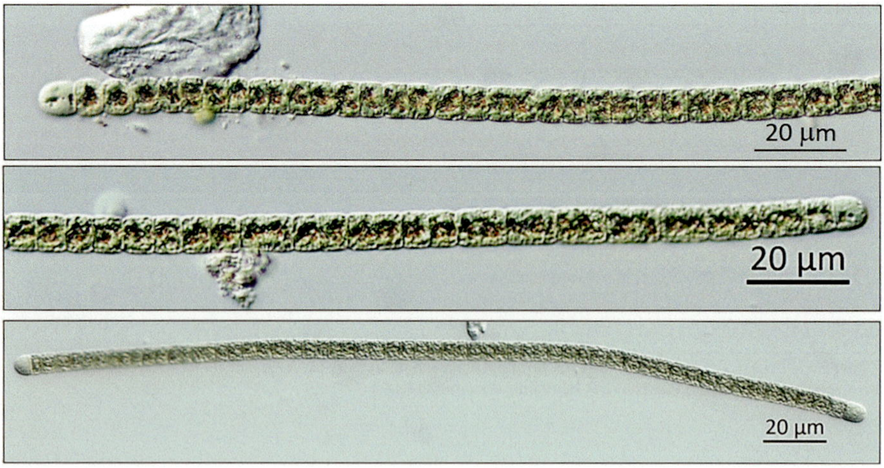

Trichodesmium iwanoffianum Plankton

Fig. 1.49 Trichodesmium

Class: Cyanophyceae
Order: Oscillatoriales
Family: Microcoleaceae

Cyanobacteria
cHAB

Tychonema

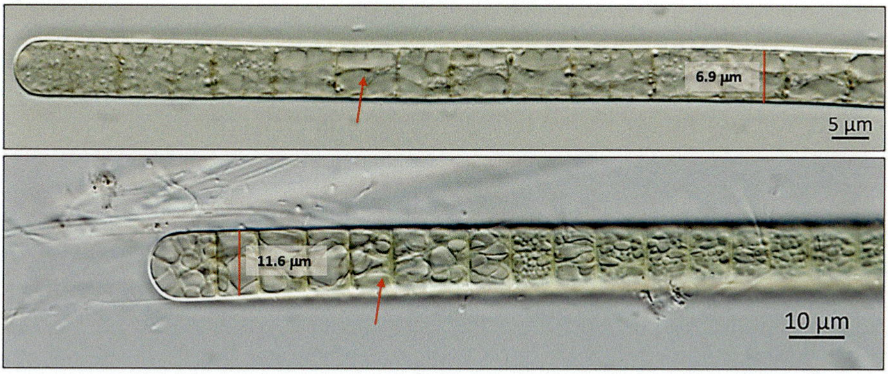

The morphological identification to species this genus is partially dependent of trichome width and color. Found as solitary trichomes or in thin benthic mats. Cells also show with visible chromatoplasm thickenings (keritomized) (at arrows). Terminal cells are rounded. Thin sheath may or may not be present.

Tychonema spp. anatoxin Metaphyton and benthic

Fig. 1.50 *Tychonema*

Class: Cyanophyceae
Order: Oscillatoriales
Family: Microcoleaceae

Cyanobacteria
cHAB

Tychonema

Trichomes 8 µm in diameter or larger. Cells "alveolar" (at arrow) and karotomized. Associated with dog mortality event in Zion National Park, Utah, USA.

Tychonema bornetii anatoxin Benthic

Fig. 1.51 *Tychonema*

Class: Cyanophyceae
Order: Oscillatoriales
Family: Microcoleaceae

Cyanobacteria
cHAB

Microcoleus

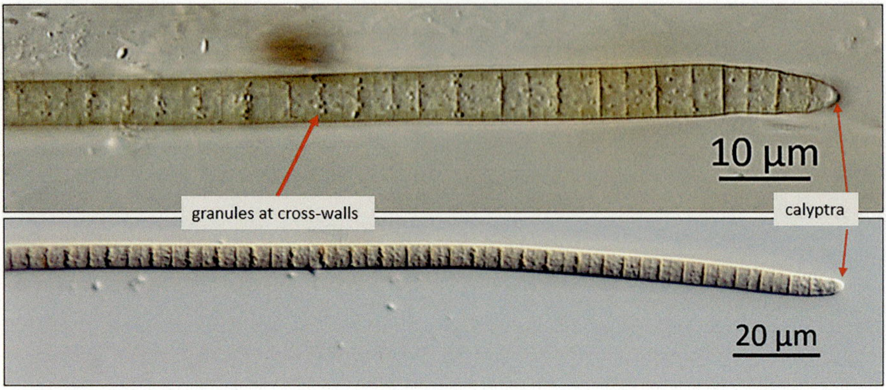

The morphology of the species with *Microcoleus* overlap with one another and other members in the Oscillatoriales. Depicted here are organisms and their characteristics that would need genetic information to properly sort them.

Microcoleus anatoxicus anatoxin Benthic

Fig. 1.52 *Microcoleus*

Class: Cyanophyceae
Order: Oscillatoriales
Family: Microcoleaceae

Cyanobacteria
cHAB

Microcoleus

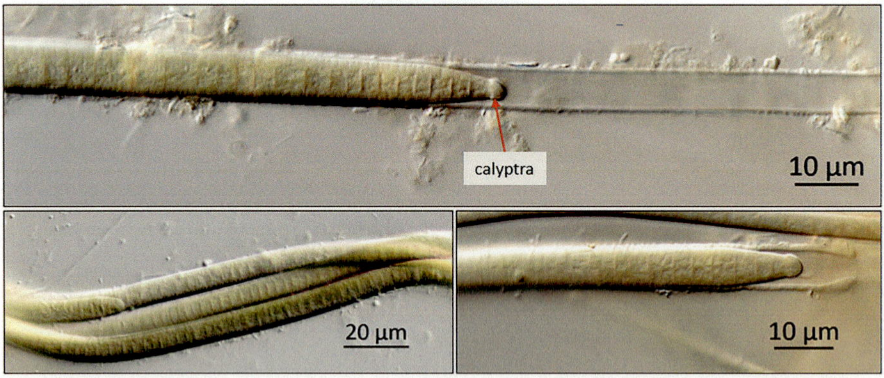

A variety of colors and morphology for can be found in this species. Often as bundles of filaments. Sheath may or not be easily observed. Cells shorter than wide. Can be found in a variety of habitats. Potential toxin producer.

Microcoleus vaginatus anatoxin Benthic

Fig. 1.53 *Microcoleus*

Class: Cyanophyceae
Order: Coleofasciculales
Family: Coleofaciculaceae

Cyanobacteria

Coleofasciculus

Coleofasciculus chthonoplastes Benthic

Fig. 1.54 *Coleofasciculus*

Class: Cyanophyceae
Order: Geitlerinematales
Family: Geitlerinemataceae

Cyanobacteria
cHAB

Geitlerinema

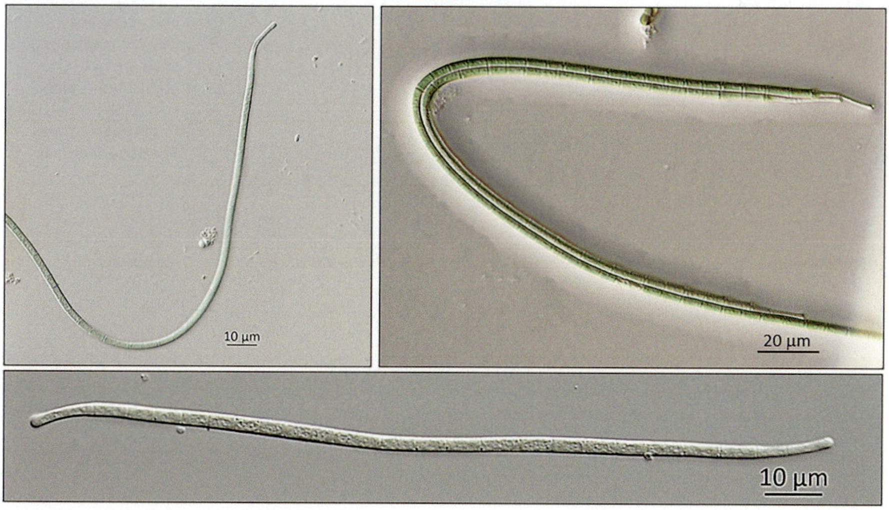

Geitlerinema splendidum microcystins Benthic

Fig. 1.55 *Geitlerinema*

Class: Cyanophyceae
Order: Coleofasciculales
Family: Wilmottiaceae

Cyanobacteria

Anagnostidinema

Anagnostidinema sp.

Benthic, metaphyton

Fig. 1.56 *Anagnostidinema*

Class: Cyanophyceae
Order: Oscillatoriales
Family: Oscillatoriaceae

Cyanobacteria
cHAB

Oscillatoria

Oscillatoria spp. are common benthic cyanobacteria, with almost 300 species that have a variety of morphologies and habitats.
Cells are shorter than wide, not constricted at the cross-walls (or slightly). The terminal end(s) of the trichomes may be bent and not have a calyptra (cap). Cross walls are not granulated. Colors vary widely from green to yellows, browns. Sheaths are absent.

Oscillatoria spp.

microcystins, anatoxin, cylindrospermopsins

Benthic, metaphyton, periphyton

Fig. 1.57 *Oscillatoria*

Class: Cyanophyceae
Order: Oscillatoriales
Family: Microcoleaceae

Cyanobacteria
cHAB

Lyngbya

Lyngbya spp. are common benthic cyanobacteria, with over 200 species that have a variety of morphologies and habitats, freshwater and marine.

The sheath is the prominent feature of this genus, along with cells are shorter than wide, not constricted at the cross-walls (or slightly).

Lyngbyatoxins, which are dermatotoxins, are associated with this genus; the most common source has been reclassified to *Microseira*.

Lyngbya spp. dermatotoxins Benthic

Fig. 1.58 *Lyngbya*

Class: Cyanophyceae
Order: Oscillatoriales
Family: Microcoleaceae

Cyanobacteria
cHAB

Limnoraphis

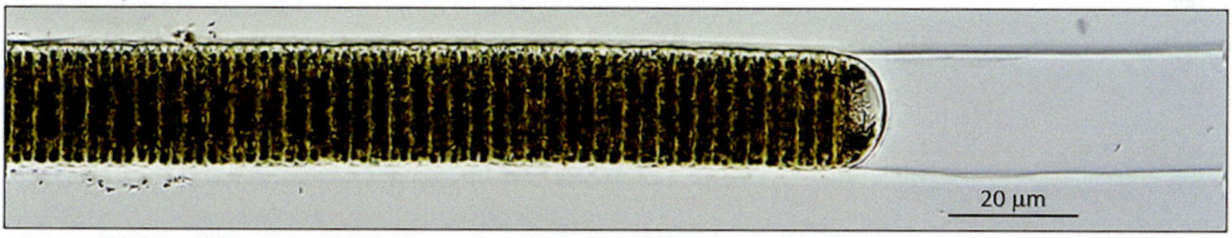

Limnoraphis birgei (filaments 19-20 µm)

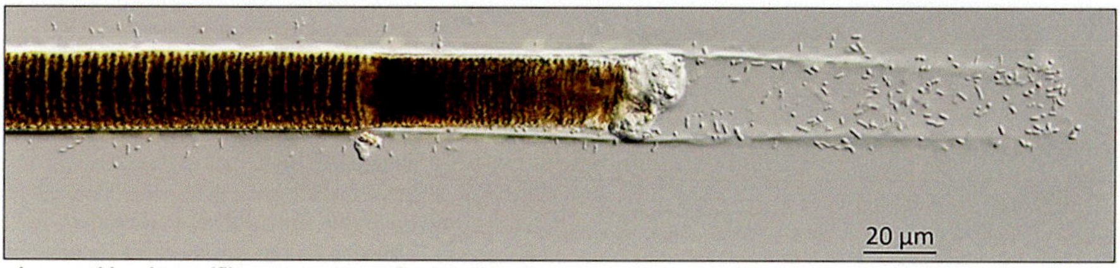

Limnoraphis robusta (filaments 20-24 µm)

Similar to *Lyngbya*, however, this is a planktonic genus, and a bloom-former make aerotopes. These species are distinguished by the width of the filament.

saxitoxins Plankton

Fig. 1.59 *Limnoraphis*

Class: Cyanophyceae
Order: Oscillatoriales
Family: Aerosakkonemataceae

Cyanobacteria
cHAB

Microseira

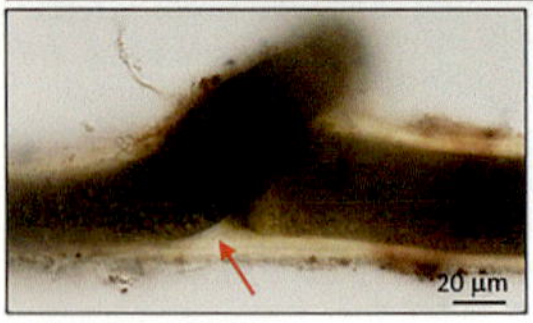

A benthic mat that can thickly cover all surfaces, often black in color. Wide cells with extensive sheath, keritomized, and rarely false branching (at arrow).

Microseira wollei saxitoxins, cylindrospermopsins Benthic

Fig. 1.60 *Microseira*

Class: Cyanophyceae
Order: Oscillatoriales
Family: Oscillatoriaceae

Cyanobacteria
cHAB

Phormidium

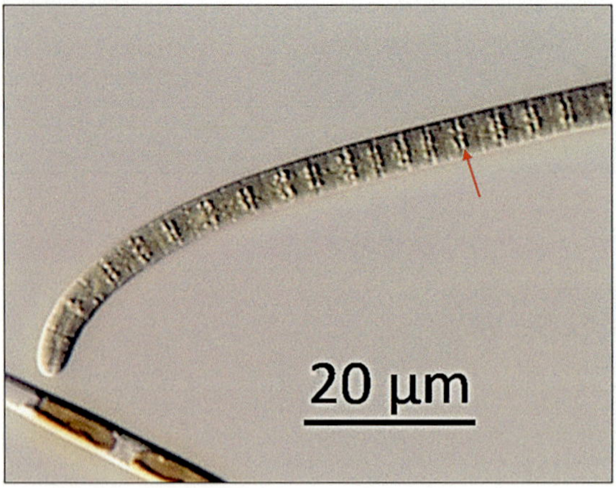

Phormidium spp. are common benthic cyanobacteria, with over 200 species that have a variety of morphologies and habitats. Several species are known to produce cyanotoxins and genetics is being used to move some species to other genera, such as *Microcoleus.*

Found as individual filaments or forming mats. Sheaths may or may not be present, have no color and are not layered and only one trichome per sheath. The cells may or may not be constricted at the cross-walls. Granules may be often found at the cross-walls (at arrow) or spread within the cells but are not essential to the identification of the genus. Terminal cells often form a curve. Color varies from green, cyan, brown, pinkish and violet. Highly motile.

Phormidium chalybeum anatoxin, microcystins Benthic, metaphyton, periphyton

Fig. 1.61 *Phormidium*

Class: Cyanophyceae
Order: Oscillatoriales
Family: Oscillatoriaceae

Cyanobacteria
cHAB

Phormidium

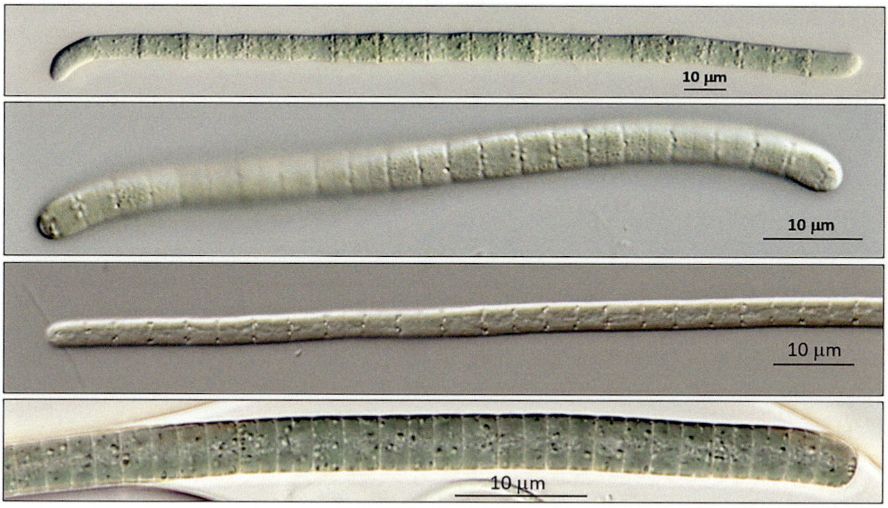

Phormidium spp. anatoxin, microcystins Benthic, metaphyton, periphyton

Fig. 1.62 *Phormidium*

Class: Cyanophyceae
Order: Oscillatoriales
Family: Oscillatoriaceae

Cyanobacteria

Phormidium

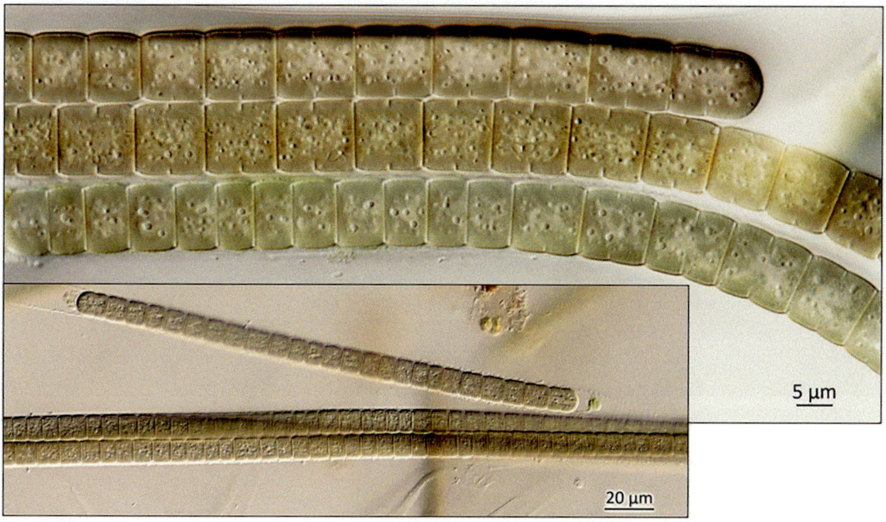

Phormidium puteale Benthic, flowing water

Fig. 1.63 *Phormidium*

Class: Cyanophyceae
Order: Oscillatoriales
Family: Oscillatoriaceae

Cyanobacteria

Phormidium

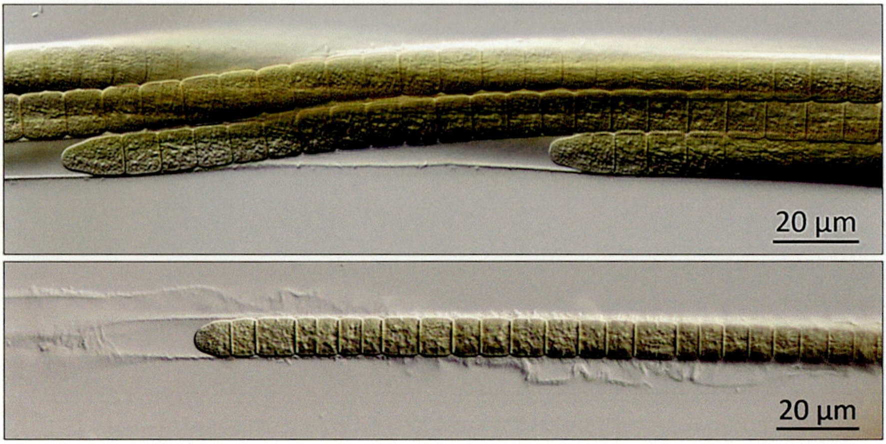

Phormidium tinctorium

Benthic, attached in flowing water, subtropical

Fig. 1.64 *Phormidium*

Class: Cyanophyceae
Order: Leptolyngbyales
Family: Leptolynbyaceae

Cyanobacteria

Planktolyngbya

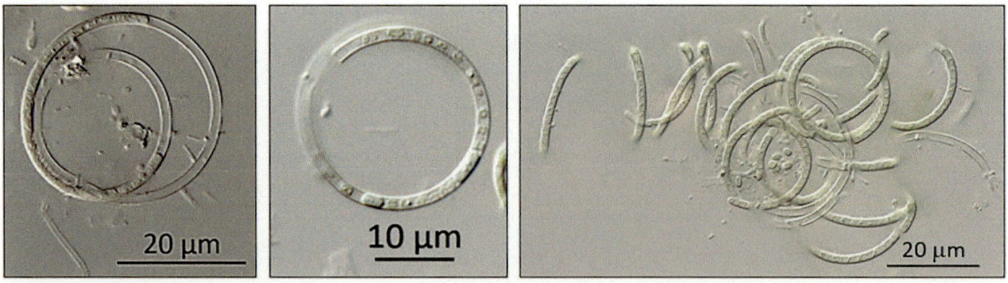

Planktolyngbya contorta

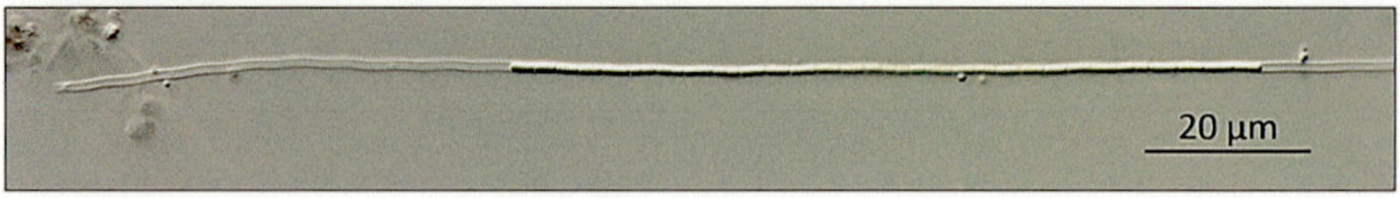

Planktolyngbya limnetica

Plankton

Fig. 1.65 *Planktolyngbya*

Class: Cyanophyceae
Order: Nostocales
Family: Aphanizomenonaceae

Cyanobacteria
cHAB

Aphanizomenon

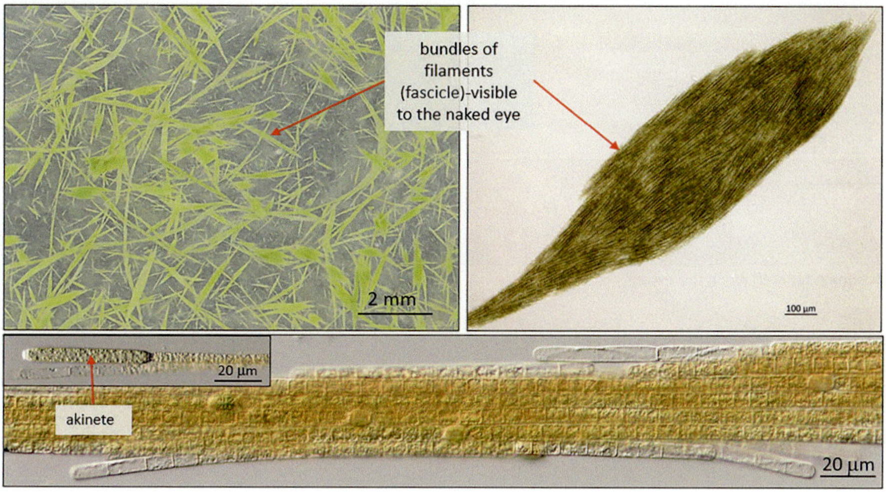

Aphanizomenon flos-aquae saxitoxins, cylindrospermopsins Plankton

Fig. 1.66 *Aphanizomenon*

Class: Cyanophyceae
Order: Nostocales
Family: Aphanizomenonaceae

Cyanobacteria
cHAB

Aphanizomenon

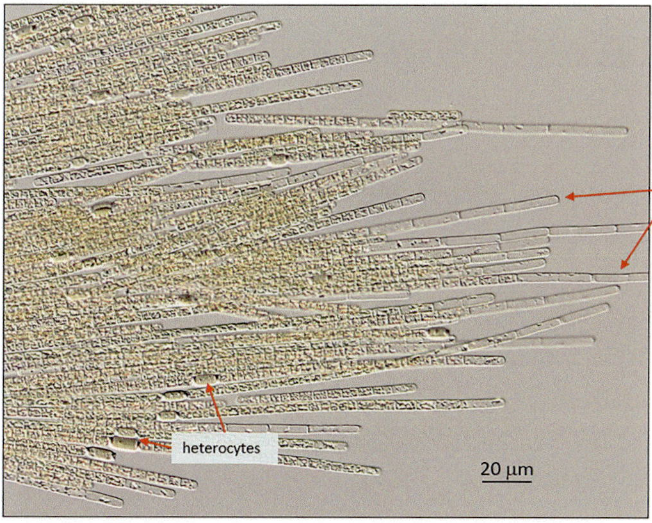

Aphanizomenon flos-aquae saxitoxins, cylindrospermopsins Plankton

Fig. 1.67 *Aphanizomenon*

Class: Cyanophyceae
Order: Nostocales
Family: Aphanizomenonaceae

Cyanobacteria
cHAB

Aphanizomenon

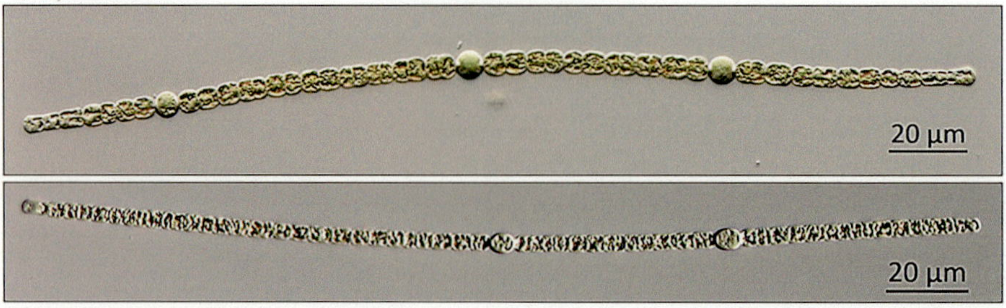

Trichome tapered along the extent

Aphanizomenon gracile | cylindrospermopsins, anatoxin | Plankton

Fig. 1.68 *Aphanizomenon*

Class: Cyanophyceae
Order: Nostocales
Family: Nodulariaceae

Cyanobacteria
cHAB

Nodularia

Nodularia spumigena | nodularin | Plankton

Fig. 1.69 *Nodularia*

Class: Cyanophyceae
Order: Nostocales
Family: Aphanizomenonaceae

Cyanobacteria
cHAB

Dolichospermum

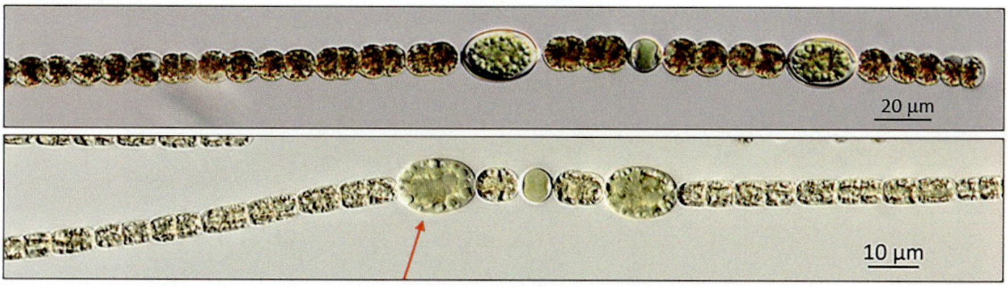

Akinete shape and position is critical to identify species

Dolichospermum planctonicum anatoxin, cylindrospermopsins Plankton

Fig. 1.70 *Dolichospermum*

Class: Cyanophyceae
Order: Nostocales
Family: Aphanizomenonaceae

Cyanobacteria
cHAB

Dolichospermum

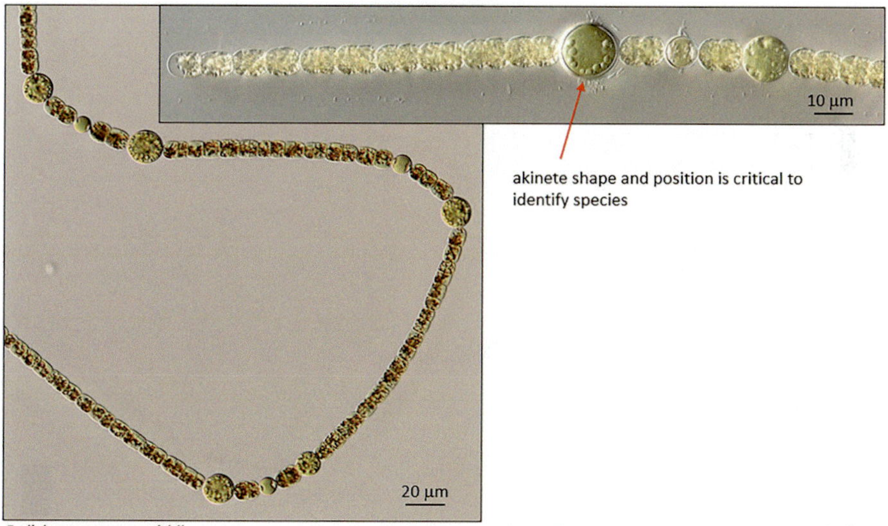

akinete shape and position is critical to identify species

Dolichospermum smithii unknown toxin profile Plankton

Fig. 1.71 *Dolichospermum*

Class: Cyanophyceae
Order: Nostocales
Family: Aphanizomenonaceae

Cyanobacteria
cHAB

Dolichospermum

Dolichospermum flos-aquae anatoxin, microcystins, guanitoxin Plankton

Fig. 1.72 *Dolichospermum*

Class: Cyanophyceae
Order: Nostocales
Family: Aphanizomenonaceae

Cyanobacteria
cHAB

Dolichospermum

Dolichospermum circinale microcystins, saxitoxins Plankton

Fig. 1.73 *Dolichospermum*

Class: Cyanophyceae
Order: Nostocales
Family: Aphanizomenonaceae

Cyanobacteria
cHAB

Dolichospermum

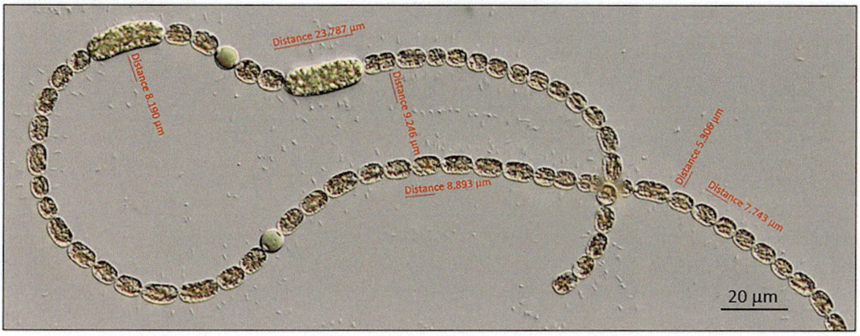

Dolichospermum sigmoideum unknown toxin profile Plankton

Fig. 1.74 *Dolichospermum*

Class: Cyanophyceae
Order: Nostocales
Family: Aphanizomenonaceae

Cyanobacteria
cHAB

Dolichospermum

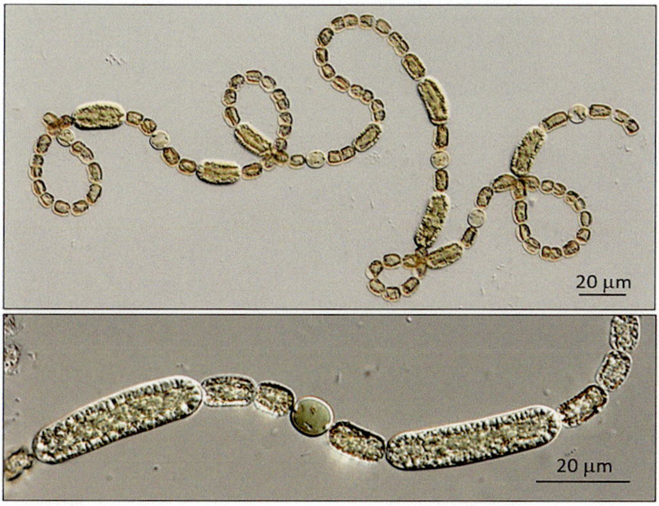

Dolichospermum berezowskii unknown toxin profile Cold water lakes Oregon, USA Plankton

Fig. 1.75 *Dolichospermum*

Class: Cyanophyceae
Order: Nostocales
Family: Aphanizomenonaceae

Cyanobacteria
cHAB

Dolichospermum

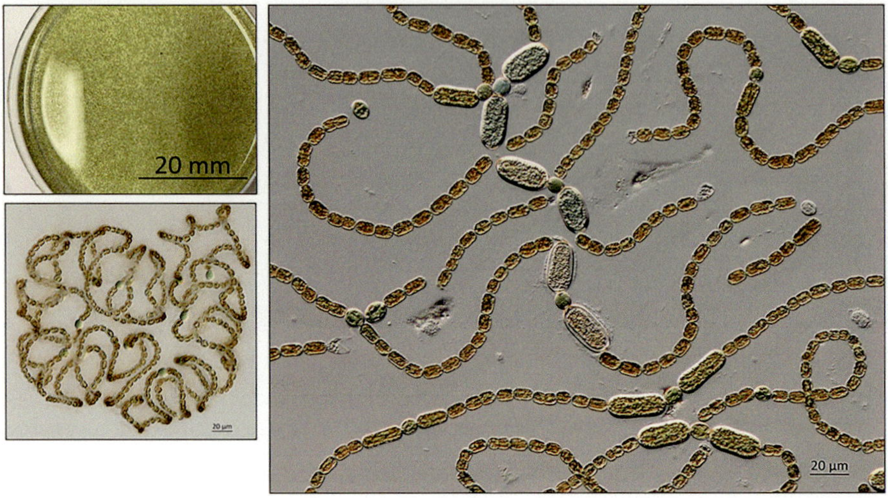

Dolichospermum lemmermannii microcystins, saxitoxins, guanitoxin Plankton

Fig. 1.76 *Dolichospermum*

Class: Cyanophyceae
Order: Nostocales
Family: Aphanizomenonaceae

Cyanobacteria
cHAB

Dolichospermum

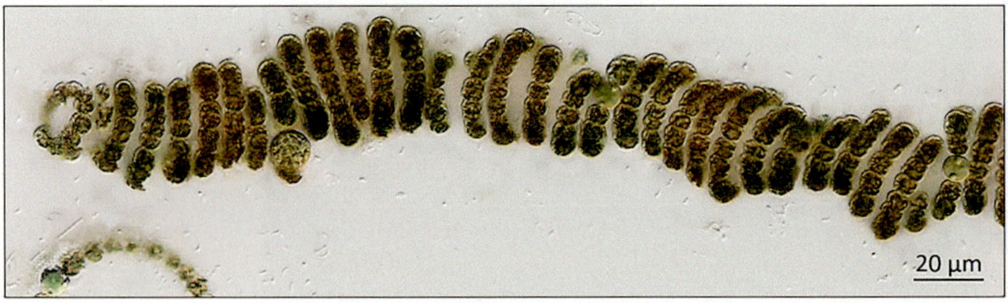

Dolichospermum compactum anabaenopeptins Plankton

Fig. 1.77 *Dolichospermum*

Class: Cyanophyceae
Order: Nostocales
Family: Aphanizomenonaceae

Cyanobacteria
cHAB

Dolichospermum

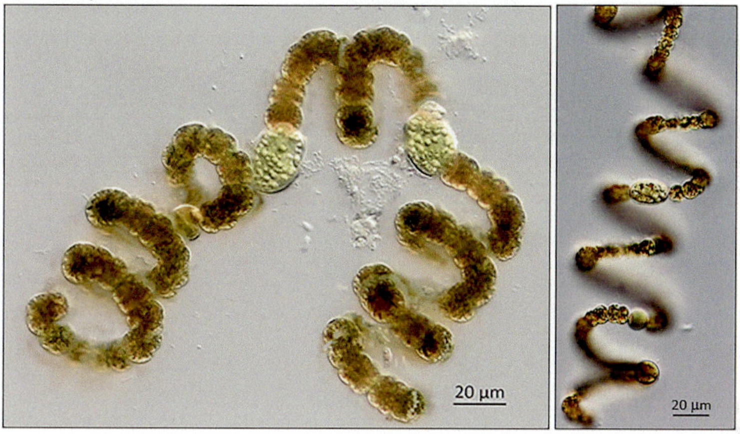

Dolichospermum crassum unknown toxin profile Plankton

Fig. 1.78 *Dolichospermum*

Class: Cyanophyceae
Order: Nostocales
Family: Aphanizomenonaceae

Cyanobacteria
cHAB

Dolichospermum

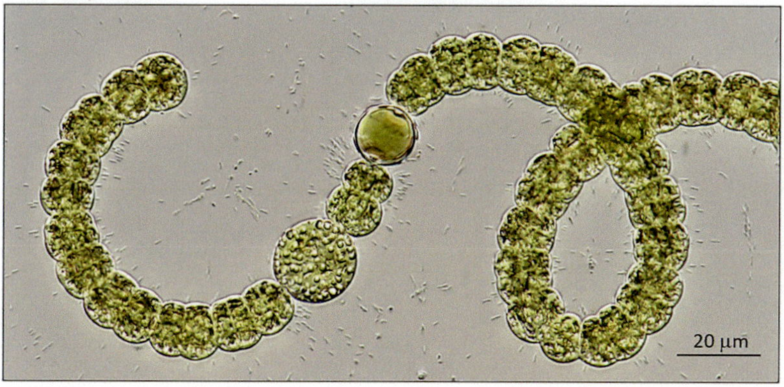

Dolichospermum mucosum unknown toxin profile Plankton

Fig. 1.79 *Dolichospermum*

Class: Cyanophyceae
Order: Nostocales
Family: Aphanizomenonaceae

Cyanobacteria

Dolichospermum

Bundles of filaments (fascicle)

Dolichospermum affine Plankton

Fig. 1.80 *Dolichospermum*

Class: Cyanophyceae
Order: Nostocales
Family: Nodulariaceae

Cyanobacteria
HAB

Anabaenopsis

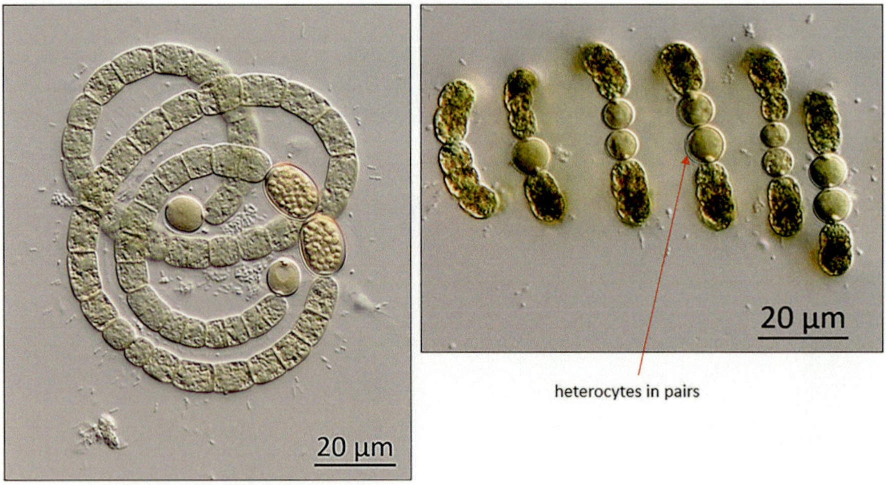

Anabaenopsis magna microcystins Plankton

Fig. 1.81 *Anabaenopsis*

Class: Cyanophyceae
Order: Nostocales
Family: Aphanizomenonaceae

Cyanobacteria
cHAB

Sphaerospermopsis (see next previous species for straight form of this genus)

Sphaerospermopsis torques-reginae guanitoxin, saxitoxin, anabaenopeptins, spumigin Plankton

Fig. 1.82 *Sphaerospermopsis*

Class: Cyanophyceae
Order: Nostocales
Family: Aphanizomenonaceae

Cyanobacteria
cHAB

Sphaerospermopsis

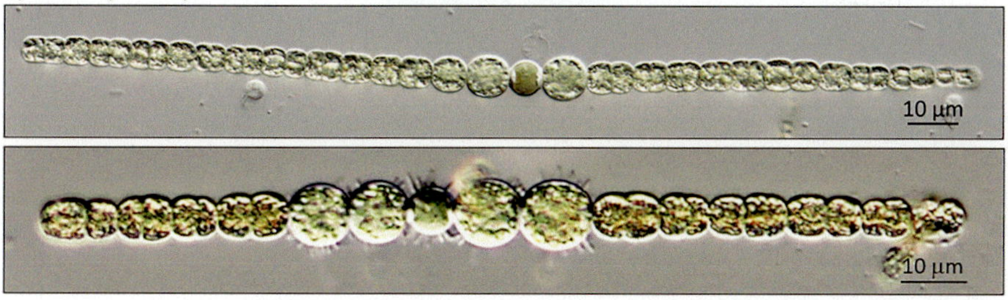

Sphaerospermopsis aphanizomenoides unknown toxin profile Plankton

Fig. 1.83 *Sphaerospermopsis*

Class: Cyanophyceae
Order: Nostocales
Family: Aphanizomenonaceae

Cyanobacteria
cHAB

Umezakia (formerly *Chrysosporum*)

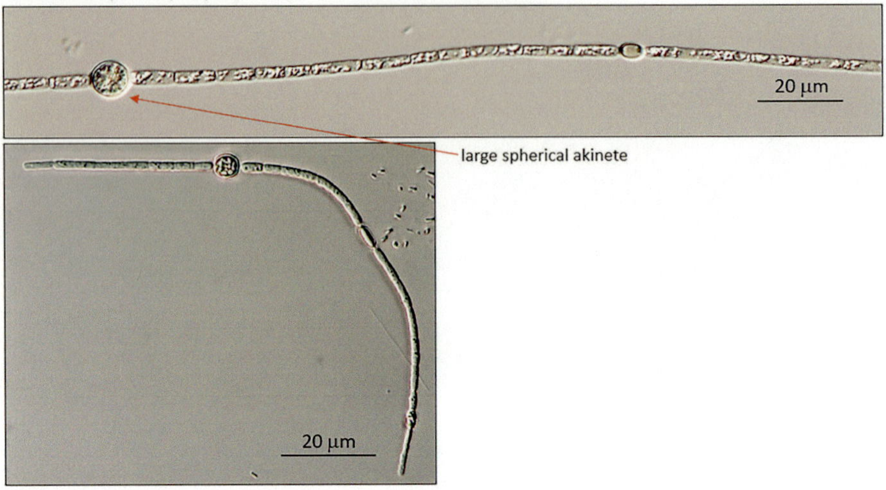

Umezakia ovalisporum anatoxin, cylindrospermopsins Plankton

Fig. 1.84 *Umezakia*

Class: Cyanophyceae
Order: Nostocales
Family: Aphanizomenonaceae

Cyanobacteria
cHAB

Raphidiopsis (formerly *Cylindrospermopsis*)

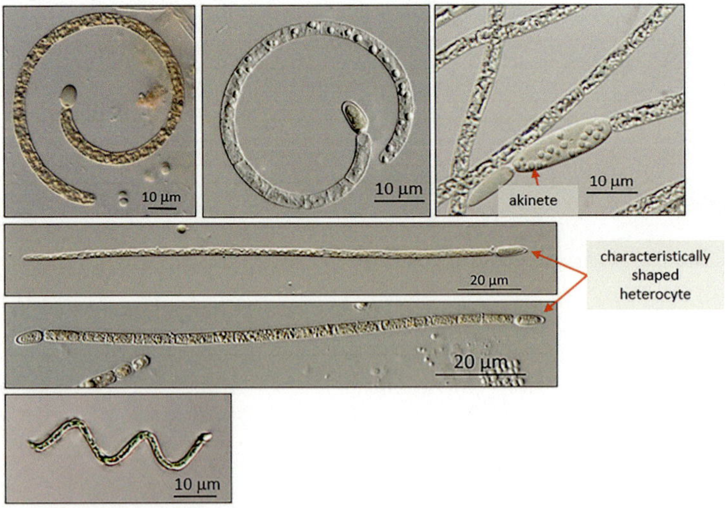

Raphidiopsis spp. anatoxin, saxitoxins, cylindrospermopsins Plankton

Fig. 1.85 *Raphidiopsis*

Class: Cyanophyceae
Order: Nostocales
Family: Aphanizomenonaceae

Cyanobacteria
cHAB

Cuspidothrix

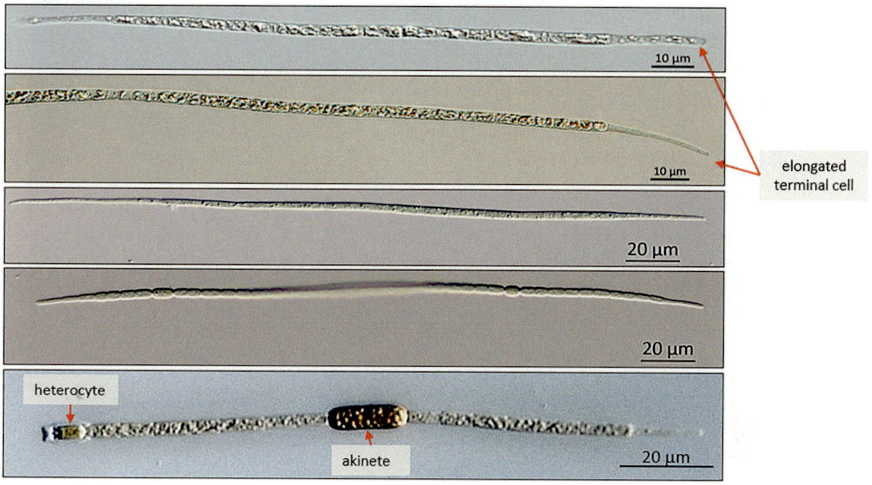

Cuspidothrix spp   anatoxin   Plankton

Fig. 1.86 *Cuspidothrix*

Class: Cyanophyceae
Order: Nostocales
Family: Aphanizomenonaceae

Cyanobacteria
cHAB

Macrospermum

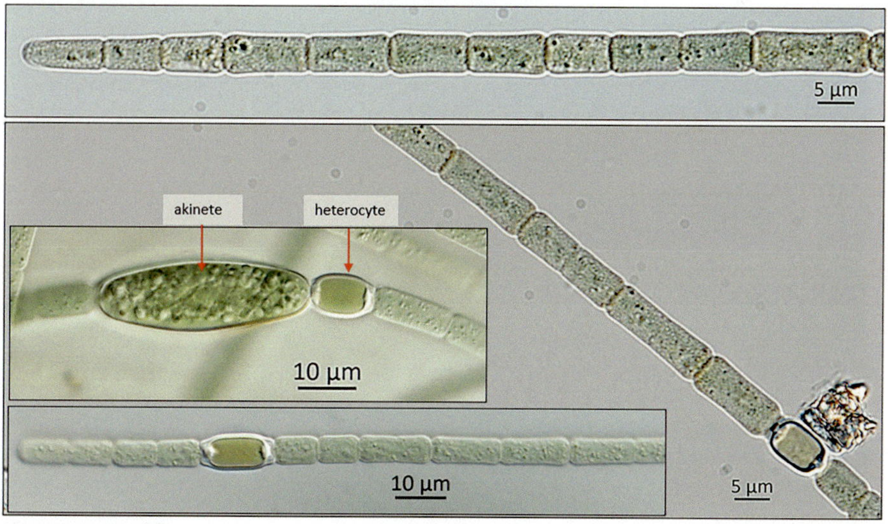

Macrospermum volzii   anatoxin   Metaphyton and plankton

Fig. 1.87 *Macrospermum*

Class: Cyanophyceae
Order: Nostocales
Family: Rivulariaceae

Cyanobacteria
cHAB

Calothrix

Calothrix spp. microcystins Benthic, attached to various substrata

Fig. 1.88 *Calothrix*

Class: Cyanophyceae
Order: Nostocales
Family: Rivulariaceae

Cyanobacteria

Dulciacalothrix

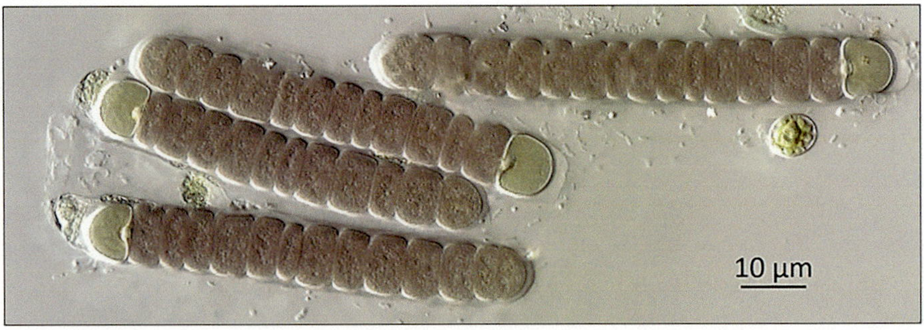

Dulcicalothrix sp Benthic, attached to various substrata

Fig. 1.89 *Dulciacalothrix*

Class: Cyanophyceae
Order: Nostocales
Family: Aphanizomenonaceae

Cyanobacteria
cHAB

Gloeotrichia

Gloeotrichia echinulata microcystins Plankton

Fig. 1.90 *Gloeotrichia*

Class: Cyanophyceae
Order: Nostocales
Family: Aphanizomenonaceae

Cyanobacteria
cHAB

Gloeotrichia

Gloeotrichia natans microcystins Plankton

Fig. 1.91 *Gloeotrichia*

Class: Cyanophyceae
Order: Nostocales
Family: Leptobasaceae

Cyanobacteria

Microchaete

Microchaete tenera Benthic

Fig. 1.92 *Microchaete*

Class: Cyanophyceae
Order: Nostocales
Family: Leptobasaceae

Cyanobacteria

Aulosira

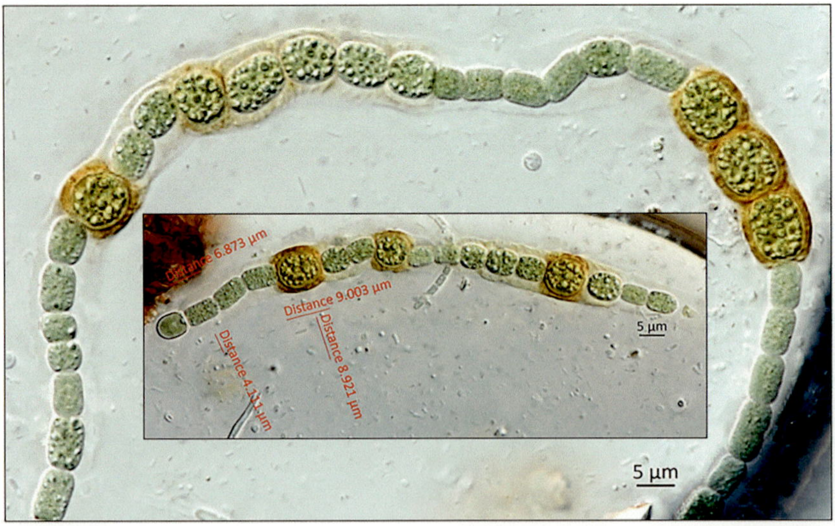

Aulosira valkanovii Flooded soils in a Cypress forest (*Taxodium*), Florida, USA

Fig. 1.93 *Aulosira*

Class: Cyanophyceae
Order: Nostocales
Family: Leptobasaceae

Cyanobacteria

Aulosira

Aulosira valkanovii Flooded soils in a Cypress forest (*Taxodium*), Florida, USA

Fig. 1.94 *Aulosira*

Class: Cyanophyceae
Order: Nostocales
Family: Nostocaceae

Cyanobacteria
cHAB

Cylindrospermum

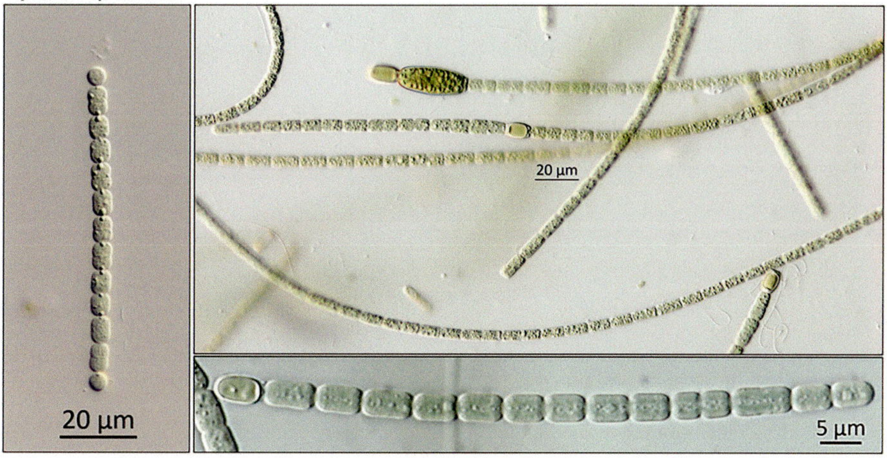

Cylindrospermum spp. anatoxin, cylindrospermopsin Benthic

Fig. 1.95 *Cylindrospermum*

Class: Cyanophyceae
Order: Nostocales
Family: Nostocaceae

Cyanobacteria
cHAB

Nostoc

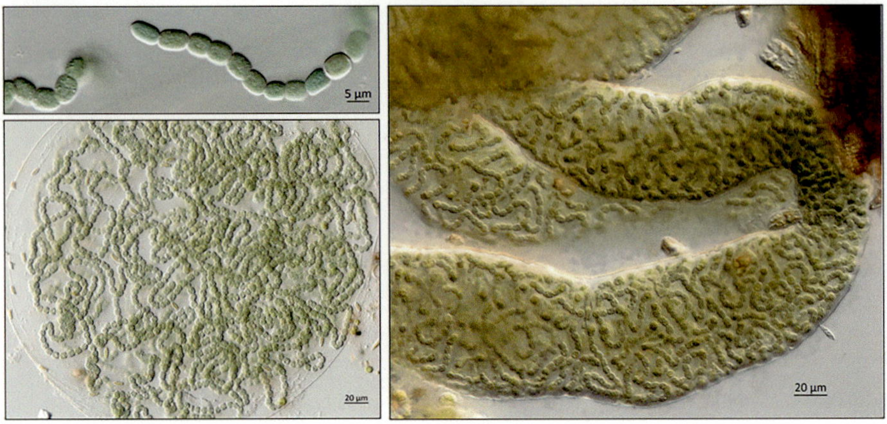

Nostoc spp. nodularin, microcystins Periphyton, metaphyton, benthic

Fig. 1.96 *Nostoc*

Class: Cyanophyceae
Order: Nostocales
Family: Nostocaceae

Cyanobacteria
cHAB

Anabaena

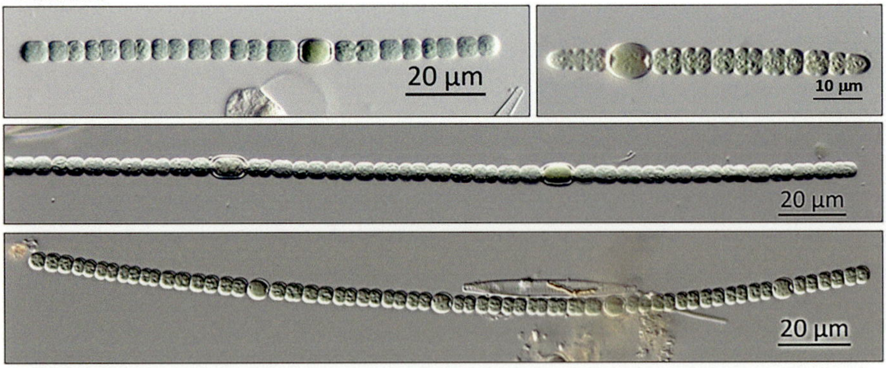

Anabaena, a name that once included many of planktonic cyanobacteria such as *Dolichospermum,* lack aerotopes which distinguishes this genus from planktonic cyanobacteria.

Anabaena spp. anatoxin, microcystins Periphyton, metaphyton, benthic

Fig. 1.97 *Anabaena*

Class: Cyanophyceae
Order: Nostocales
Family: Nostocaceae

Cyanobacteria
cHAB

Anabaena

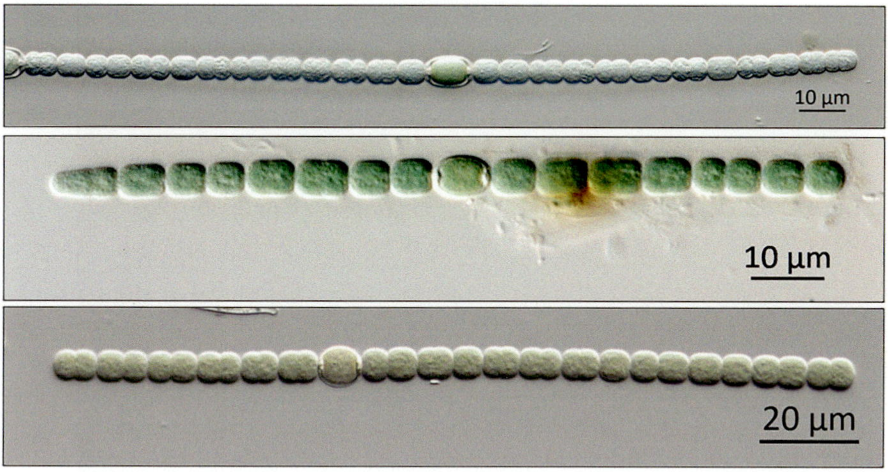

Anabaena spp. | Anatoxin, microcystin | Periphyton, metaphyton, benthic

Fig. 1.98 *Anabaena*

Class: Cyanophyceae
Order: Nostocales
Family: Nostocaceae

Cyanobacteria

Anabaena

Anabaena fuscovaginata | Calcareous marshes, metaphyton, benthic

Fig. 1.99 *Anabaena*

Class: Cyanophyceae
Order: Nostocales
Family: Leptobasaceae

Cyanobacteria

Fortiea

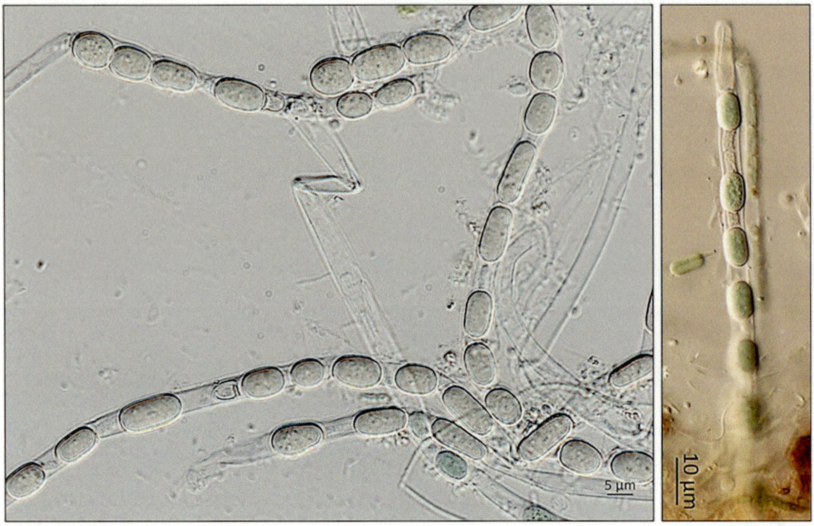

Fortiea contorta Metaphyton, benthic

Fig. 1.100 *Fortiea*

Class: Cyanophyceae
Order: Nostocales
Family: Aphanizomenonaceae

Cyanobacteria

Wollea

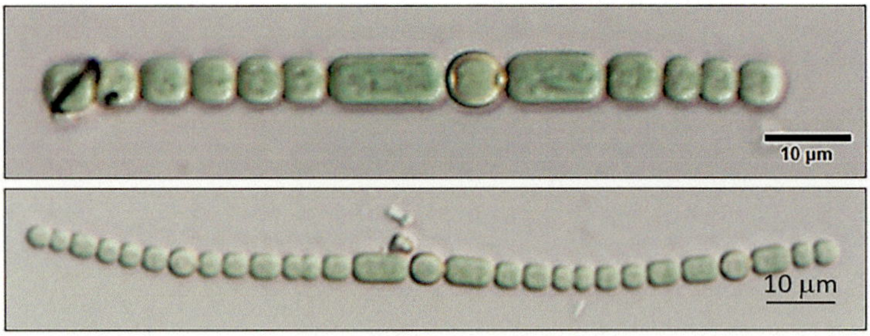

Wollea sp. Benthic

Fig. 1.101 *Wollea*

Class: Cyanophyceae
Order: Nostocales
Family: Hapalosiphonaceae

Cyanobacteria
cHABs

Hapalosiphon

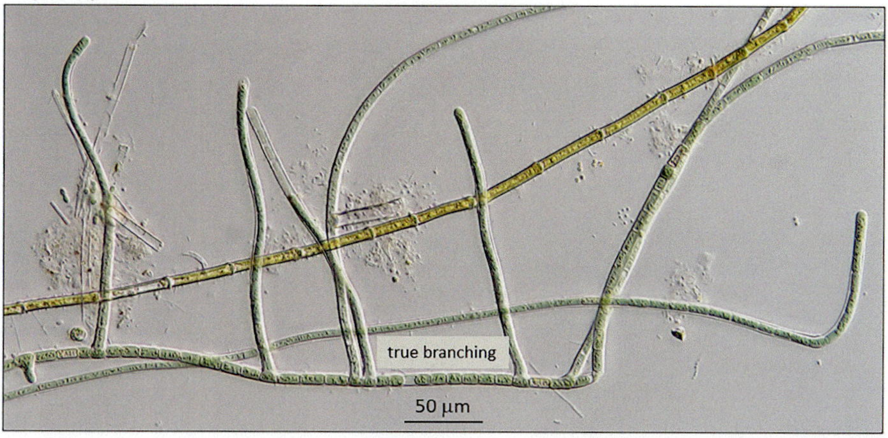

Hapalosiphon hibernicus microcystins Metaphyton, benthic

Fig. 1.102 *Hapalosiphon*

Class: Cyanophyceae
Order: Nostocales
Family: Hapalosiphonaceae

Cyanobacteria

Nostochopsis

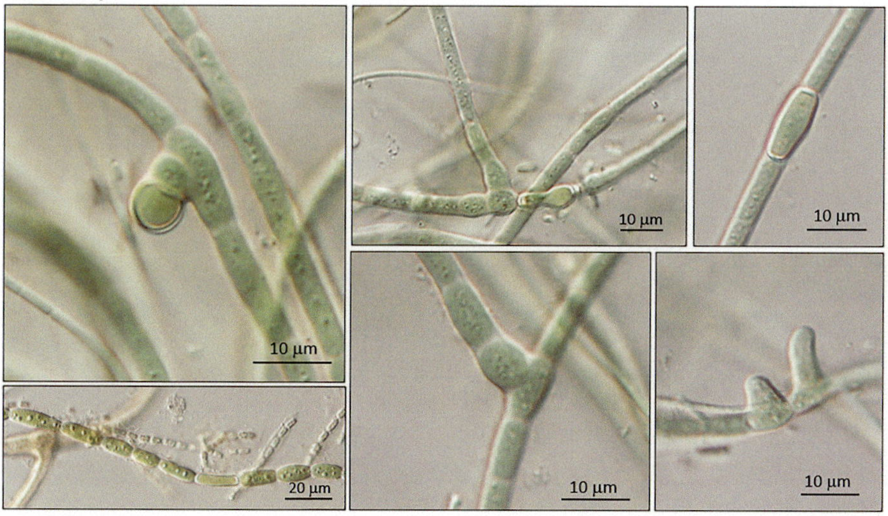

Nostochopsis sp. Metaphyton, benthic

Fig. 1.103 *Nostochopsis*

Class: Cyanophyceae
Order: Notocales
Family: Sytonemataceae

Cyanobacteria

Brasilonema

Brasilonema bambusicola

Splash zone

Fig. 1.104 *Brasilonema*

Class: Cyanophyceae
Order: Nostocales
Family: Tolypothrichaceae

Cyanobacteria

Tolypothrix

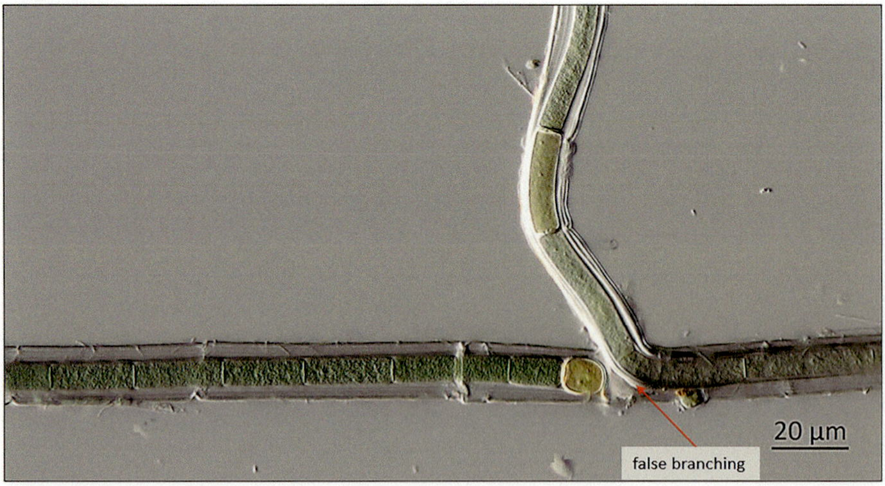

Tolypothrix sp.

Metaphyton, benthic

Fig. 1.105 *Tolypothrix*

Class: Cyanophyceae
Order: Nostocales
Family: Tolypothrichaceae

Cyanobacteria

Tolypothrix

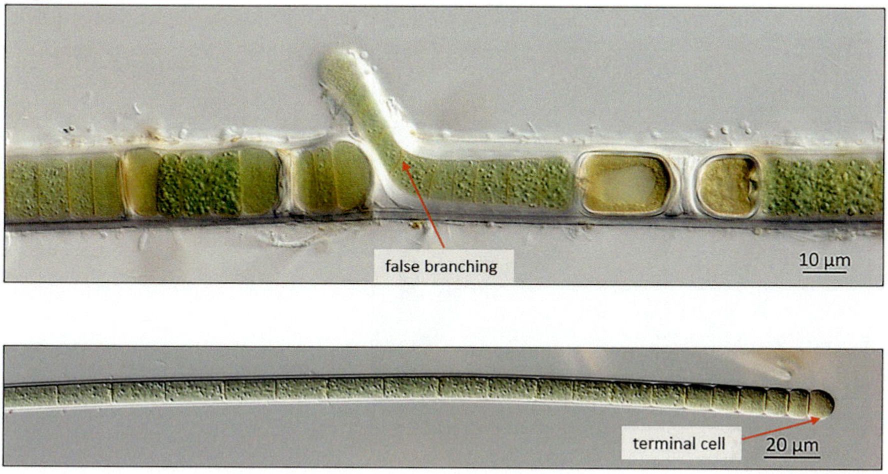

Tolypothrix sp.

Metaphyton, benthic

Fig. 1.106 *Tolypothrix*

Class: Cyanophyceae
Order: Nostocales
Family: Scytonemataceae

Cyanobacteria
cHAB

Scytonema

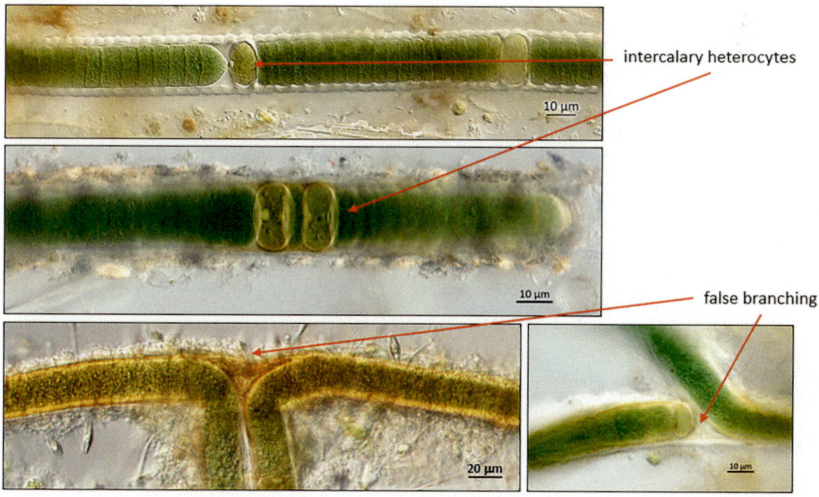

Scytonema spp.

Saxitoxin

Periphytic, benthic

Fig. 1.107 *Scytonema*

Class: Cyanophyceae
Order: Nostocales
Family: Scytonemataceae

Cyanobacteria

Chakia

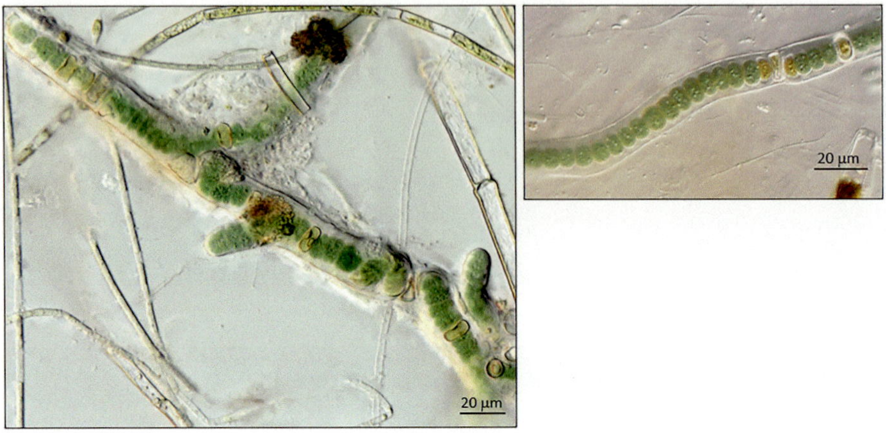

Chakia cubana Periphytic, benthic

Fig. 1.108 *Chakia*

Class: Cyanophyceae
Order: Nostocales
Family: Scytonemataceae

Cyanobacteria

Iphinoë

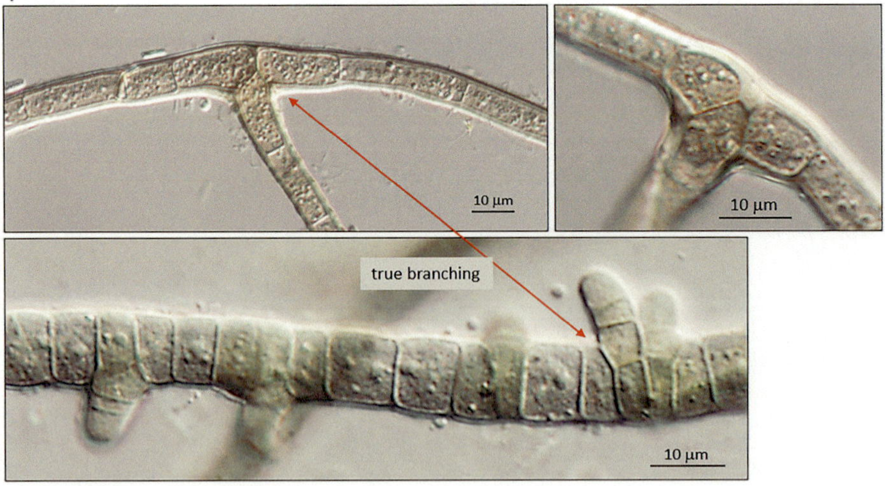

Iphinoë spelaea Periphytic, benthic

Fig. 1.109 *Iphinoë*

Class: Cyanophyceae
Order: Nostocales
Family: Stigonemataceae

Cyanobacteria

Stigonema

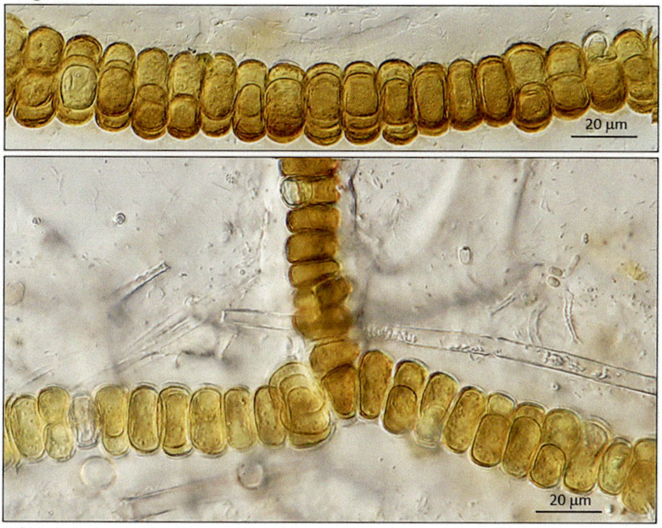

This genus is multiseriate, when cells form more than a single strand, rather attach to one another in more than one plane.

Stigonema sp.

Periphytic, benthic

Fig. 1.110 *Stigonema*

Class: Cyanophyceae
Order: Nostocales
Family: Stigonemataceae

Cyanobacteria

Stigonema

Stigonema turfaceum

Periphytic, benthic

Fig. 1.111 *Stigonema*

Suggested Reading

Anagnostidis, K., & Komárek, J. (1985). Modern approach to the classification system of cyanophytes. 1 – Introduction. *Archiv für Hydrobiologie Supplement, 71*(Algological Studies 38–39), 291–302.

Anagnostidis, K., & Komárek, J. (1988). Modern approach to the classification system of cyanophytes. 3. Oscillatoriales. *Archiv für Hydrobiologie, Supplement, 80*(Algological Studies 50–53 (1–4)), 327–472.

Anagnostidis, K., & Komárek, J. (1990). Modern approach to the classification system of cyanophytes. 5. Stigonematales. *Archiv für Hydrobiologie Supplement, 86*(Algological Studies 59), 1–73.

Bohunická, M., Mareš, J., Hrouzek, P., Urajová, P., Lukeš, M., Šmarda, J., Komárek, J., Gaysina, L. A., & Strunecký, O. (2015). A combined morphological, ultrastructural, molecular, and biochemical study of the peculiar family Gomontiellaceae (Oscillatoriales) reveals a new cylindrospermopsin-producing clade of cyanobacteria. *Journal of Phycology, 51*(6), 1040–1054.

Bohunická, M., Johansen, J. R., Villanueva, C. D., Mareš, J., Štenclová, L., Becerra-Absalón, I., Hauer, T., & Kaštovský, J. (2024). Revision of the pantropical genus *Brasilonema* (Nostocales, Cyanobacteria), with the description of 24 species new to science. *Fottea, 24*(2), 137–184.

Johansen, J. R., Bohunická, M., Lukesová, A., Hrcková, K., Vaccarino, M. A., & Chesarino, N. M. (2014). Morphological and molecular characterization within 26 strains of the genus *Cylindrospermum* (Nostocaceae, Cyanobacteria), with descriptions of three new species. *Journal of Phycology, 50*(1), 187–202.

Komárek, J. (1985). Do all cyanophytes have a cosmopolitan distribution? Survey of the freshwater cyanophyte flora of Cuba. *Archiv für Hydrobiologie (Supplement), 71*(Algological Studies 38–39), 359–386.

Komárek, J. (2003). Coccoid and colonial cyanobacteria. In J. D. Wehr & R. G. Sheath (Eds.), *Freshwater Algae of North America* (Ecology and classification) (pp. 59–116). Academic.

Komárek, J. (2008). The cyanobacterial genus *Macrospermum. Fottea, 8*(1), 79–86.

Komárek, J. (2013). Süßwasserflora von Mitteleuropa. In *Cyanoprokaryota: Heterocystous Genera (Nostocales, Stigonematales)* (Vol. 19/3, pp. [i]–xviii, [1]-1130). Springer Spektrum.

Komárek, J., & Anagnostidis, K. (1986). Modern approach to the classification system of cyanophytes, 2 - Chroococcales. *Archiv für Hydrobiologie Supplement, 73/2*(Algological Studies 43), 157–226.

Komárek, J., & Anagnostidis, K. (1989). Modern approach to the classification system of Cyanophytes 4 - Nostocales. *Algological Studies, 56*, 247–345.

Komárek, J., & Anagnostidis, K. (1999). Cyanoprokaryota. 1. Chroococcales. In H. Ettl, G. Gärtner, H. Heynig, & D. Mollenhauer (Eds.), *Süßwasserflora von Mitteleuropa. Begründet von A. Pascher. Band 19/1* (pp. 1–548). Spektrum/Akademischer Verlag.

Komárek, J., & Anagnostidis, K. (2005). *Süßwasserflora von Mitteleuropa. Cyanoprokaryota. Oscillatoriales* (Vol. 19/2 pp. 759, 1010 figures). Elsevier Spektrum Akademischer Verlag.

Komárek, J., & Johansen, J. R. (2015). Filamentous Cyanobacteria. In J. D. Wehr, R. G. Sheath, & J. P. Kociolek (Eds.), *Freshwater Algae of North America* (Ecology and classification) (2nd ed., pp. 135–235). Academic.

Komárek, J., & Kastovsky, J. (2003). Coincidences of structural and molecular characters in evolutionary lines of cyanobacteria. *Algological Studies/Archiv für Hydrobiologie, Supplement Volumes, 109*, 305–325.

Komárek, J., Cepák, V., Kastovský, J., & Sulek, J. (2004). What are the cyanobacterial genera *Cyanothece* and *Cyanobacterium*? Contribution to the combined molecular and phenotype taxonomic evaluation of cyanobacterial diversity. *Algological Studies/Archiv für Hydrobiologie, Supplement Volumes, 113*, 1–36.

Komarek, J., Zapomelova, E., Smarda, J., Kopecky, J., Rejmankova, E., Woodhouse, J., Neilan, A. B., & Komarkova, J. (2013). Polyphasic evaluation of *Limnoraphis robusta* a water-bloom forming cyanobacterium from Lake Atitlán, Guatemala, with a description of *Limnoraphis* gen. nov. *Fottea, 13*(1), 39–52.

Komárek, J., Kaštovský, J., Mares, J., & Johansen, J. R. (2014). Taxonomic classification of cyanoprokaryotes (cyanobacterial genera) 2014, using a polyphasic approach. *Preslia, 86*, 295–335.

Komárková, J., Zapomelová, E., & Komárek, J. (2013). *Chakia* (cyanobacteria), a new heterocytous genus from Belizean marshes identified on the basis of the 16S rRNA gene. *Fottea, 13*(2), 227–233.

Lukesová, A., Johansen, J. R., Martin, M. P., & Casamatta, D. A. (2009). *Aulosira bohemensis* sp. nov.: Further phylogenetic uncertainty at the base of the Nostocales (Cyanobacteria). *Phycologia, 48*(2), 118–129.

Mares, J., Kaštovský, J., McGregor, G., & Komárek, J. (2008). The little known pseudofilamentous cyanobacterium *Wolskyella* (Synechococcales). *Nova Hedwigia, 87*, 221–230.

Rajaniemi, P., Komárek, J., Hoffmann, L., Hrouzek, P., Kastocská, K., & Sivonen, K. (2005). Taxonomic consequences from the combined molecular and phenotype evaluation of selected *anabaena* and *Aphanizomenon* strains. *Algological Studies (Cyanobacterial Research 6), 117*, 371–391.

Rosen, B.H. & Mareš, J. (2016). Catalog of microscopic organisms of the Everglades. Part 1: The Cyanobacteria. *U.S. Geological Survey Scientific Investigations Report* 2016–1114.

Siegesmund, M. A., Johansen, J. R., Karsten, U., & Friedl, T. (2008). *Coleofasciculus* gen. nov. (Cyanobacteria): Morphological and molecular criteria for revision of the genus *Microcoleus* Gomont. *Journal of Phycology, 44*, 1572–1585.

Simic, S. B., Komárek, J., & Dordevic, N. B. (2014). The confirmation of the genus *Glaucospira* (cyanobacteria) and the occurrence of *Glaucospira laxissima* (G. S. West) comb. nova in Serbia. *Cryptogamie Algologie, 35*(3), 259–267.

Strunecky, O., Komárek, J., Johansen, J., Lukesová, A., & Elster, J. (2013). Molecular and morphological criteria for revision of the genus *Microcoleus* (Oscillatoriales, cyanobacteria). *Journal of Phycology, 49*(6), 1167–1180.

Strunecky, O., Bohunická, M., Johansen, J. R., Capková, K., Raabová, L., Dvorák, P., & Komárek, J. (2017). A revision of the genus *Geitlerinema* and a description of the genus *Anagnostidinema* gen. nov. (Oscillatoriophycidae, Cyanobacteria). *Fottea, 17*(1), 114–126.

Wacklin, P., Hoffmann, L., & Komárek, J. (2009). Nomenclatural validation of the genetically revised cyanobacterial genus *Dolichospermum* (Ralfs ex Bornet et Flahault) comb. nova. *Fottea, 9*(1), 59–64.

Green Algae

2

The green algae are a diverse group of photosynthetic organisms that thrive in aquatic and some terrestrial environments. They belong to the kingdom Plantae, and one branch of this group led to the land plants. They are primarily characterized by their green pigmentation, due to the presence of chlorophyll *a* and chlorophyll *b*. As with all algae, these pigments have an active role in photosynthesis, converting light energy into chemical energy and contributing significantly to the primary productivity of freshwater ecosystems. In addition, green algae also have complementary accessory pigments, the carotenoids, that expand the usable wavelengths for photosynthesis and pass that energy on to chlorophyll *a*. Carotenoids are also involved in photoprotection and can impart their coloration on the cells observed.

The cell wall of green algae is primarily composed of cellulose, providing structural integrity and rigidity to the cells. In addition to cellulose, many species possess a matrix of hemicellulose and pectin that contributes to the wall's overall composition, making it flexible yet strong. It also provides a good substrate for epiphytic algae. There are a great variety of cell shapes and sizes, ranging from unicellular forms to multicellular filaments. Many species also secrete a mucilaginous sheath, which aids in protection against desiccation and enhances nutrient absorption.

In terms of motility, many green algae are motile, because they have a flagellum. Flagellar motility is prominent in unicellular species, such as *Chlamydomonas*, which possess two anterior flagella that facilitate phototaxis, the ability to swim toward (or away) from light. Some species only have a motile stage when producing gametes for reproduction. Several genera, like the multicellular forms or colonies, are completely nonmotile, relying on water currents for dispersal and nutrient acquisition.

Green algae inhabit a wide range of environments from ponds and lakes to streams, rivers, and wetlands. They can be found both as planktonic organisms floating in the water column and as benthic forms attached to various substrates like rocks, sediments, or aquatic vegetation. The adaptability of green algae allows them to thrive in diverse conditions, including low-light environments, nutrient-rich waters, and habitats with fluctuating water levels. They play a crucial ecological role, serving as a primary food source for a myriad of aquatic organisms and contributing to the overall health of freshwater ecosystems (Figs. 2.1 to 2.147).

B. H. Rosen, *Color Atlas of Freshwater Algae*,
https://doi.org/10.1007/978-3-031-93122-2_2

Fig. 2.1 *Chlamydomonas*

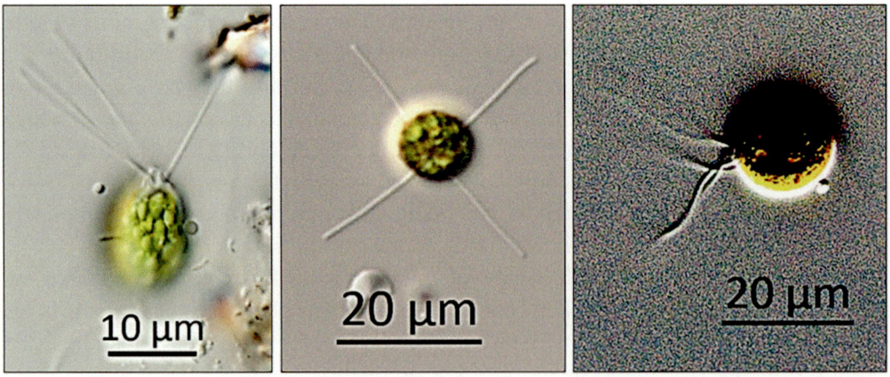

Fig. 2.2 *Carteria*

Class: Chlorophyceae
Order: Vovlvocales
Family: Dunaliellaceae

Chlorophyta

Spermatozopsis

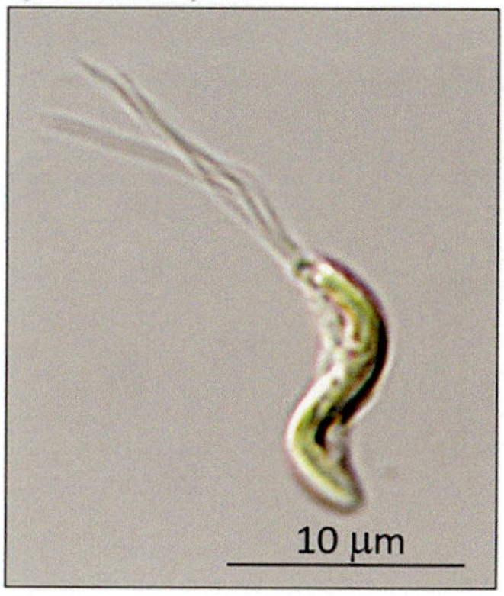

4 flagella

Spermatozopsis exsultans

Plankton

Fig. 2.3 *Spermatozopsis*

Class: Chlorophyceae
Order: Vovlvocales
Family: Palmellopsidaceae

Chlorophyta

Chlamydocapsa

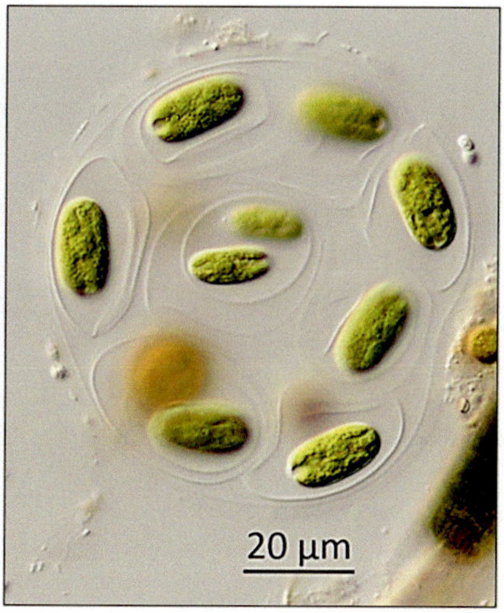

Chlamydocapsa spp.

Epiphyton, metaphyton

Fig. 2.4 *Chlamydocapsa*

Class: Chlorophyceae
Order: Vovlvocales
Family: Palmellospsidaceae

Chlorophyta

Asterococcus

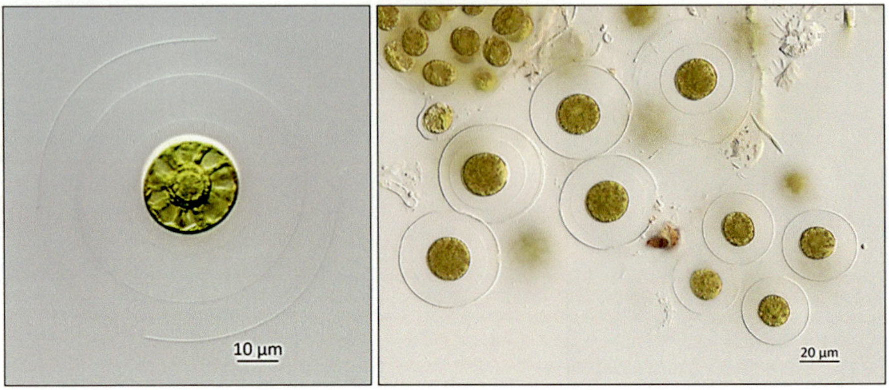

Asterococcus superbus Epiphyton, metaphyton

Fig. 2.5 *Asterococcus*

Class: Chlorophyceae
Order: Vovlvocales
Family: Chlamydomonadaceae

Chlorophyta

Desmatractum

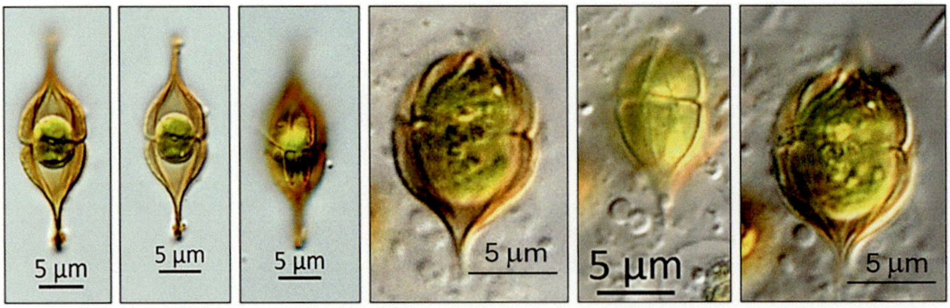

Note: the taxonomic Order of this genus is uncertain

Desmatractum bipyramidatum Epiphyton, metaphyton

Fig. 2.6 *Desmatractum*

Class: Chlorophyceae
Order: Vovlvocales
Family: Haematococcaceae

Chlorophyta

Haematococcus

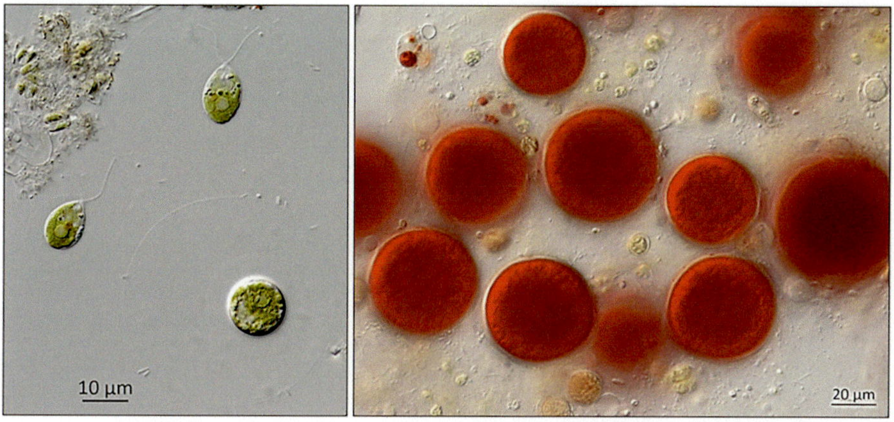

2 flagella resting cysts

Haematococcus lacustris Plankton

Fig. 2.7 *Haematococcus*

Class: Chlorophyceae
Order: Volvocales
Family: Phacotaceae

Chlorophyta

Pteromonas

Pteromonas aculeata Plankton

Fig. 2.8 *Pteromonas*

Class: Chlorophyceae
Order: Vovlvocales
Family: Phacotaceae

Chlorophyta

Pteromonas

Pteromonas angulosa Plankton

Fig. 2.9 *Pteromonas*

Class: Chlorophyceae
Order: Vovlvocales
Family: Chlamydomonadaceae

Chlorophyta

Lobomonas

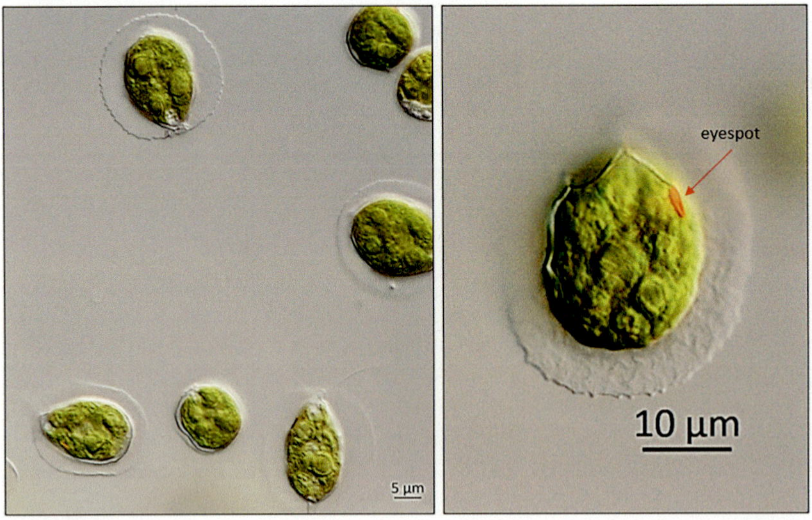

Lobomonas sp. Plankton

Fig. 2.10 *Lobomonas*

Class: Chlorophyceae
Order: Vovlvocales
Family: Phacotaceae

Chlorophyta

Phacotus

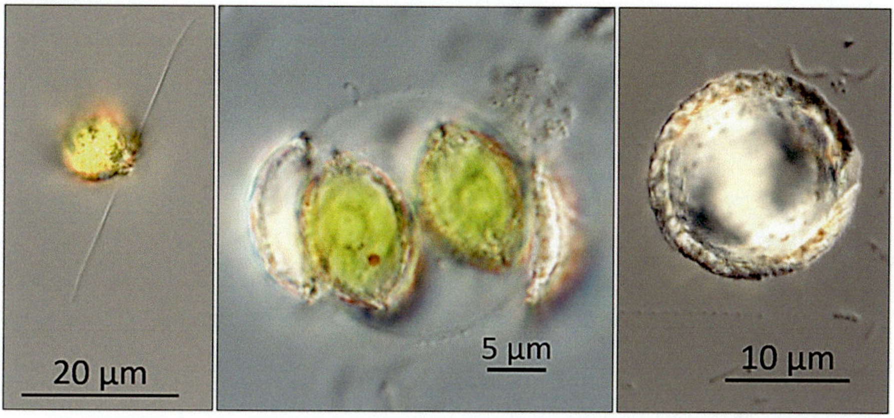

Phacotus lenticularis

Plankton

Fig. 2.11 *Phacotus*

Class: Chlorophyceae
Order: Vovlvocales
Family: Phacotaceae

Chlorophyta

Hemitoma

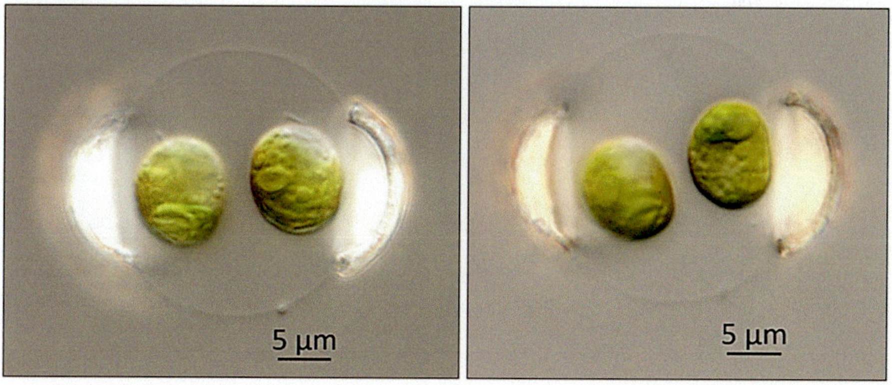

Hemitoma maeandrocystis

Plankton

Fig. 2.12 *Hemitoma*

Class: Chlorophyceae
Order: Vovlvocales
Family: Tetrabaenaceae

Chlorophyta

Tetrabaena

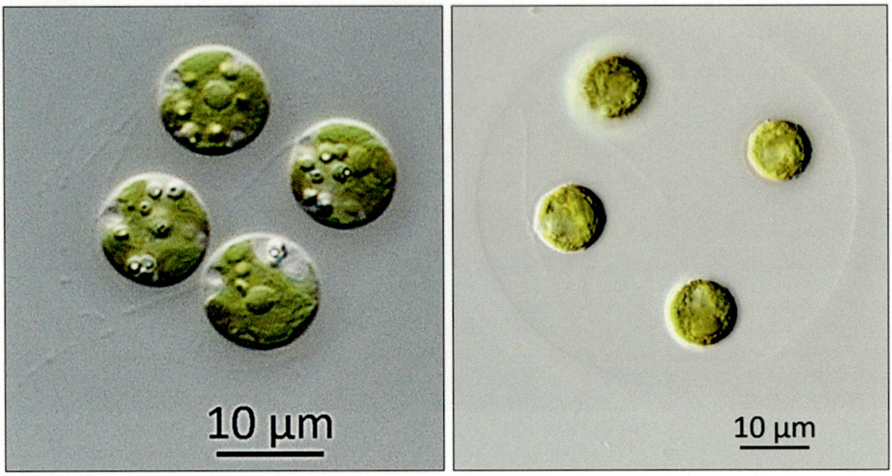

Tetrabaena socialis

Fig. 2.13 *Tetrabaena*

Class: Chlorophyceae
Order: Vovlvocales
Family: Goniaceae

Chlorophyta

Gonium

Gonium pectorale

Plankton

Fig. 2.14 *Gonium*

Class: Chlorophyceae
Order: Vovlvocales
Family: Volvocaceae

Chlorophyta

Pandorina

Pandorina morum

Plankton

Fig. 2.15 *Pandorina*

Class: Chlorophyceae
Order: Vovlvocales
Family: Volvocaceae

Chlorophyta

Eudorina

Eudorina elegans

Plankton

Fig. 2.16 *Eudorina*

Class: Chlorophyceae
Order: Vovlvocales
Family: Volvocaceae

Chlorophyta

Eudorina

Eudorina elegans

Fig. 2.17 *Eudorina*

Class: Chlorophyceae
Order: Vovlvocales
Family: Volvocaceae

Chlorophyta

Platydorina

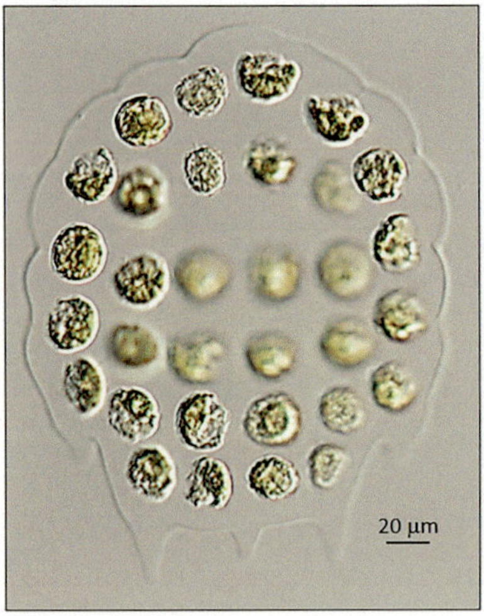

Platydorina caudata

Pleodorina

Pleodorina californica

Plankton

Fig. 2.18 *Platydorina*

Class: Chlorophyceae
Order: Vovlvocales
Family: Volvocaceae

Chlorophyta

Volvox

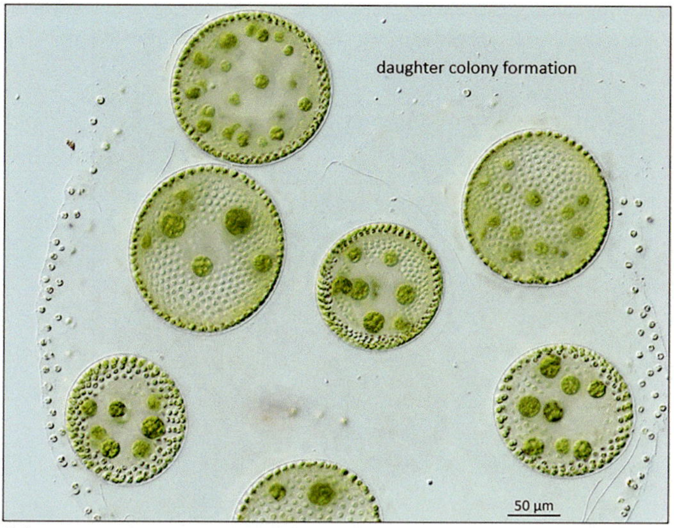

Volvox sp.

Plankton

Fig. 2.19 *Volvox*

Class: Chlorophyceae
Order: Vovlvocales
Family: Tetrasporaceae

Chlorophyta

Tetraspora

Tetraspora gelatinosa

Fig. 2.20 *Tetraspora*

Class: Chlorophyceae
Order: Vovlvocales
Family: Chlorococcaceae

Chlorophyta

Chlorococcum

Chlorococcum sp.

Fig. 2.21 *Chlorococcum*

Class: Chlorophyceae
Order: Sphaeropleales
Family: Characiaceae

Chlorophyta

Characium

Characium sp.

Characium strictum

Epiphyton

Fig. 2.22 *Characium*

Class: Ulvophyceae
Order: Ignatiales
Family: Ignatiaceae

Chlorophyta

Pseudocharacium

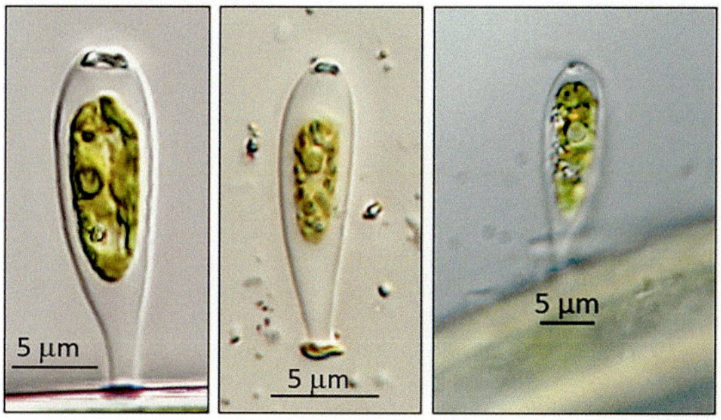

Pseudocharacium obtusum

Epiphyton

Fig. 2.23 *Pseudocharacium*

Class: Chlorophyceae
Order: Sphaeropleales
Family: Radiococcaceae

Chlorophyta

Radiococcus

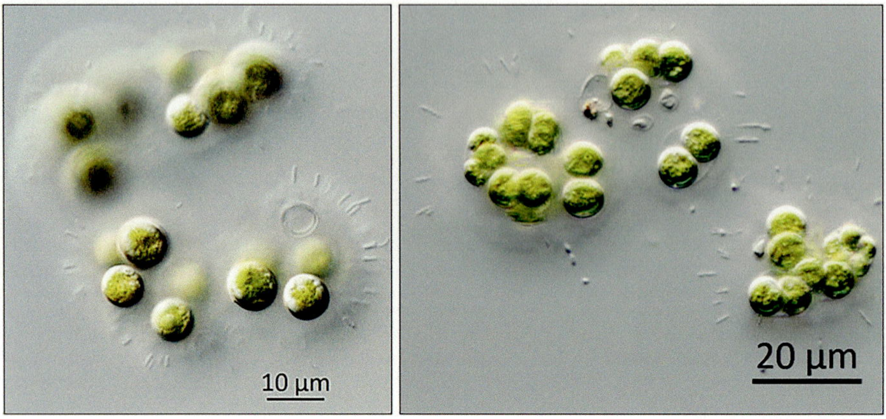

Radiococcus nimbatus

Fig. 2.24 *Radiococcus*

Class: Chlorophyceae
Order: Chaetophorales
Family: Chaetophoraceae

Chlorophyta

Stigeoclonium

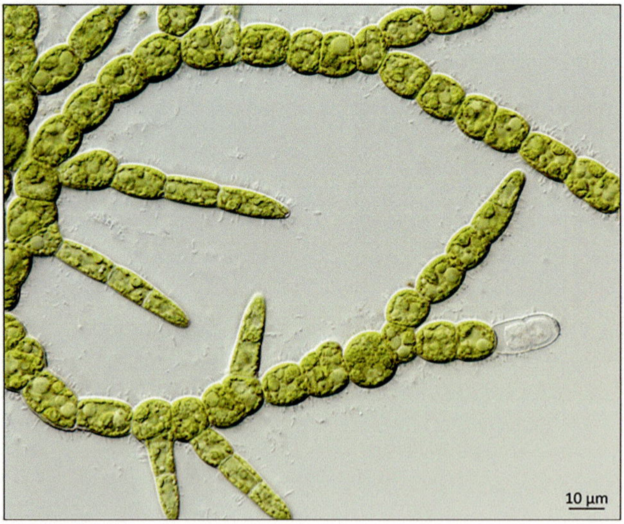

Stigeoclonium sp.

Epiphyton, epilithon

Fig. 2.25 *Stigeoclonium*

Class: Chlorophyceae
Order: Chaetophorales
Family: Chaetophoraceae

Chlorophyta

Chaetophoropsis

Chaetophoropsis pisiformis

Epiphyton, epilithic

Fig. 2.26 *Chaetophoropsis*

Class: Chlorophyceae
Order: Chaetophorales
Family: Chaetophoraceae

Chlorophyta

Chaetophoropsis

Chaetophoropsis pisiformis

Epiphyton, epilithon

Fig. 2.27 *Chaetophoropsis*

Class: Chlorophyceae
Order: Chaetophorales
Family: Chaetophoraceae

Chlorophyta

Draparnaldia

Draparnaldia mutabilis

Epiphyton, epilithic

Fig. 2.28 *Draparnaldia*

Class: Chlorophyceae
Order: Chaetophorales
Family: Chaetophoraceae

Chlorophyta

Draparnaldia

Draparnaldia mutabilis

Epiphyton, epilithon

Fig. 2.29 *Draparnaldia*

Class: Chlorophyceae
Order: Oedogoniales
Family: Oedogoniaceae

Chlorophyta

Oedogonium

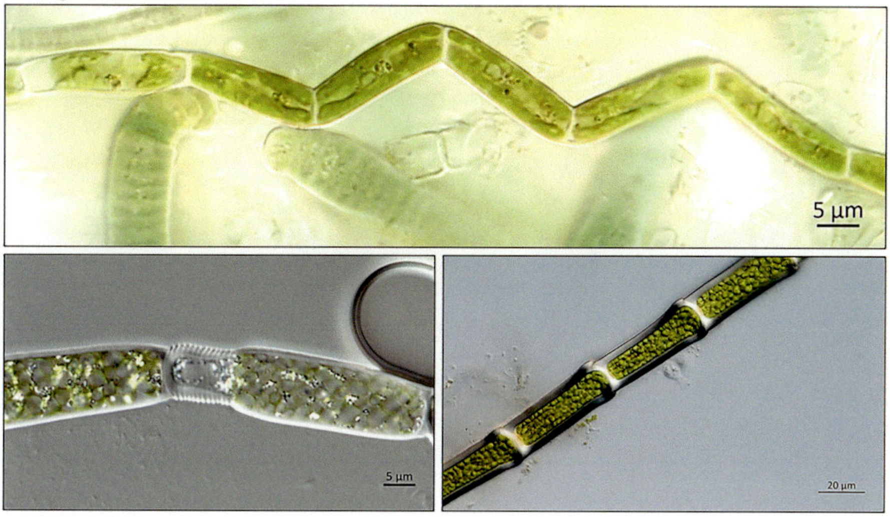

Oedogonium spp.

Epiphyton, epilithon

Fig. 2.30 *Oedogonium*

Class: Chlorophyceae
Order: Oedogoniales
Family: Oedogoniaceae

Chlorophyta

Oedogonium

Oedogonium spp.

Epiphyton, epilithon

Fig. 2.31 *Oedogonium*

Class: Chlorophyceae
Order: Oedogoniales
Family: Oedogoniaceae

Chlorophyta

Oedogonium

zoospores

Oedogonium sp.

Epiphyton, epilithon

Fig. 2.32 *Oedogonium*

Class: Chlorophyceae
Order: Oedogoniales
Family: Oedogoniaceae

Chlorophyta

Bulbochaete

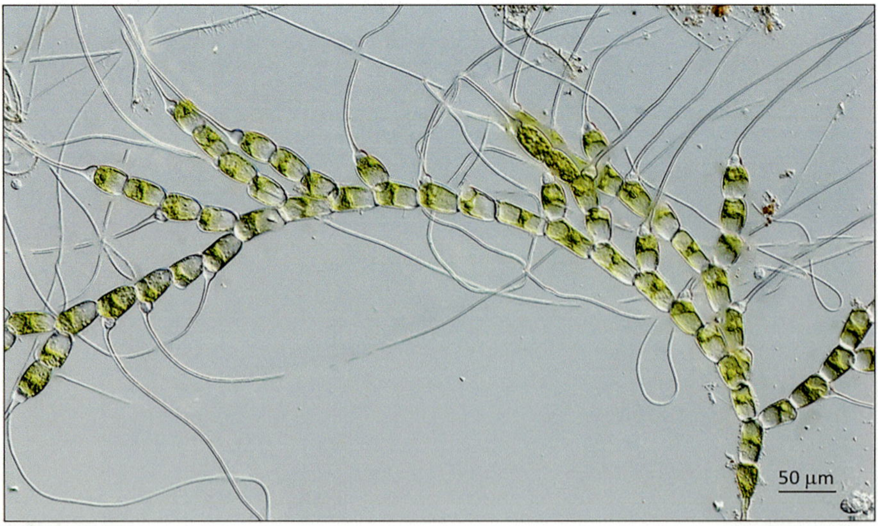

Bulbochaete spp.

Epiphyton, epilithon

Fig. 2.33 *Bulbochaete*

Class: Chlorophyceae
Order: Oedogoniales
Family: Oedogoniaceae

Chlorophyta

Bulbochaete

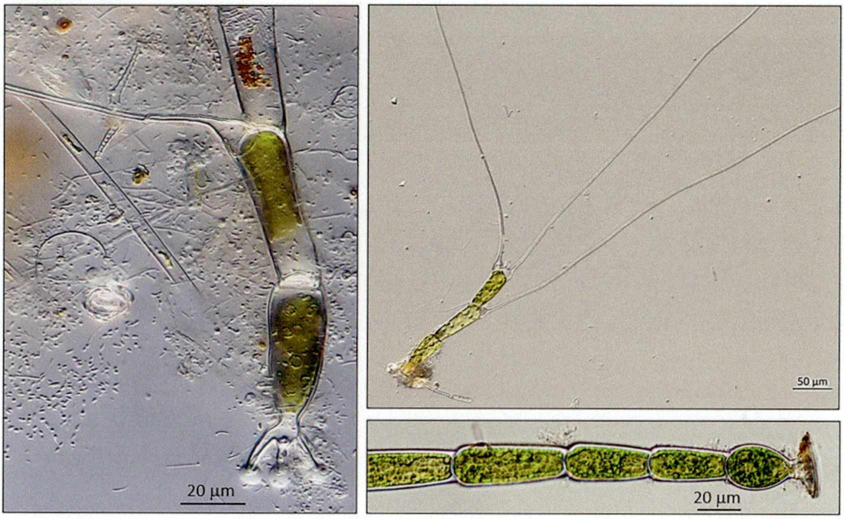

Bulbochaete sp.

Epiphyton, epilithon

Fig. 2.34 *Bulbochaete*

Class: Chlorophyceae
Order: Sphaeropleales
Family: Sphaeropleaceae

Chlorophyta

Radiofilum

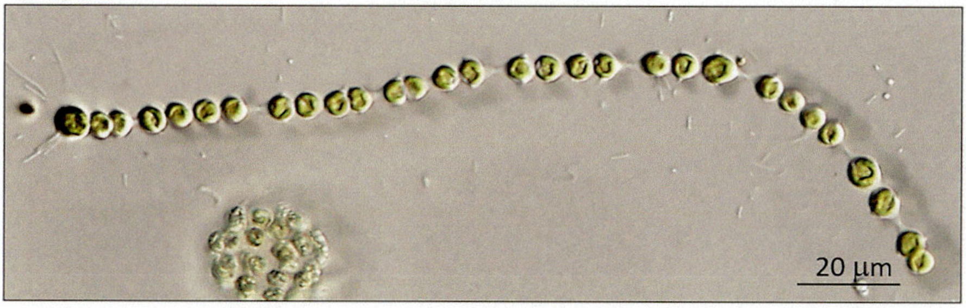

Radiofilum conjunctivum

Fig. 2.35 *Radiofilum*

Class: Chlorophyceae
Order: Sphaeropleales
Family: Microsporaceae

Chlorophyta

Microspora

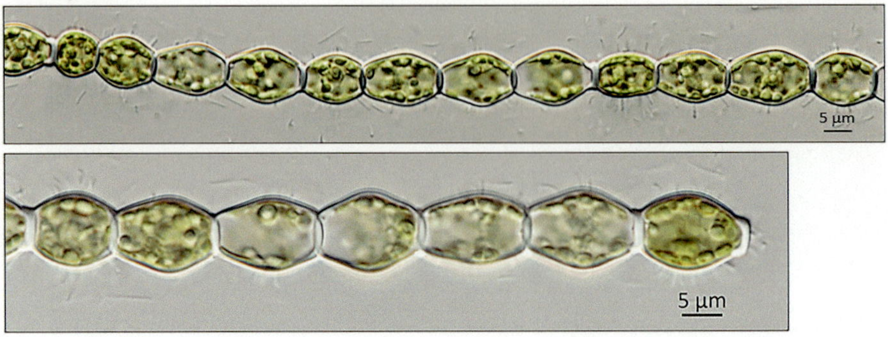

Microspora tumidula

Metaphyton

Fig. 2.36 *Microspora*

Class: Trebouxiophyceae
Order: Trebouxiophyceae *ordo incertae sedis*
Family: Trebouxiophyceae *incertae sedis*

Chlorophyta

Chlorolobion

Chlorolobion obtusum

Metaphyton

Fig. 2.37 *Chlorolobion*

Class: Chlorophyceae
Order: Sphaeropleales
Family: Selenastraceae

Chlorophyta

Ankistrodesmus

Ankistrodesmus spiralis

Ankistrodesmus sigmoideus

Metaphyton

Fig. 2.38 *Ankistrodesmus*

Class: Chlorophyceae
Order: Sphaeropleales
Family: Selenastraceae

Chlorophyta

Quadrigula

Quadrigula korsikovii

Metaphyton, plankton

Fig. 2.39 *Quadrigula*

Class: Chlorophyceae
Order: Sphaeropleales
Family: Selenastraceae

Chlorophyta

Kirchneriella

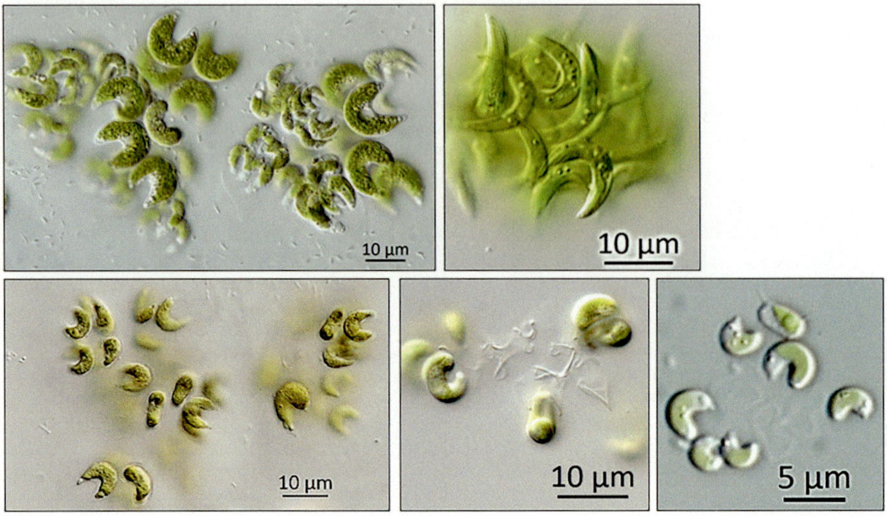

Kirchneriella spp. Metaphyton, plankton

Fig. 2.40 *Kirchneriella*

Class: Chlorophyceae
Order: Sphaeropleales
Family: Scenedesmaceae

Chlorophyta

Tetrallantos

Tetrallantos lagerheimii Metaphyton, plankton

Fig. 2.41 *Tetrallantos*

Class: Chlorophyceae
Order: Sphaeropleales
Family: Radiococcaceae

Chlorophyta

Gloeocystis

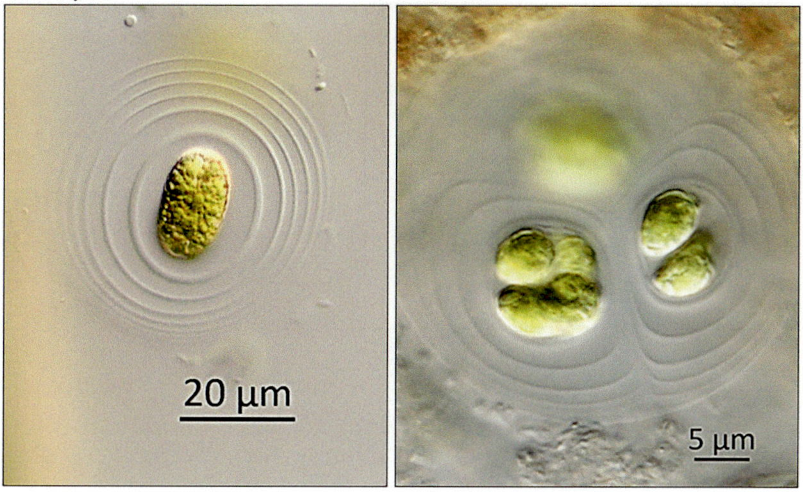

Gloeocystis botryoides

Metaphyton, epiphyton

Fig. 2.42 *Gloeocystis*

Class: Chlorophyceae
Order: Sphaeropleales
Family: Radiococcaceae

Chlorophyta

Palmodictyon

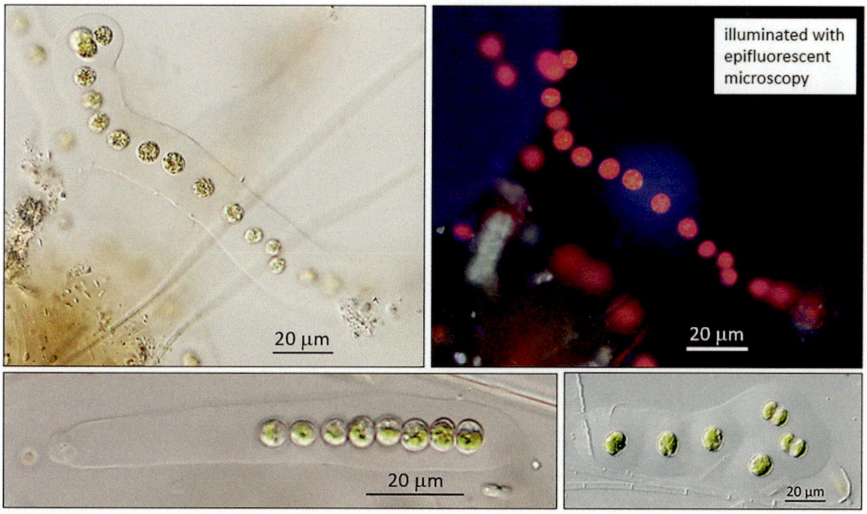

Palmodictyon viride

Metaphyton, epiphyton

Fig. 2.43 *Palmodictyon*

Class: Chlorophyceae
Order: Sphaeropleales
Family: Radiococcaceae

Chlorophyta

Palmodictyon

Palmodictyon lobatum Metaphyton, epiphyton

Fig. 2.44 *Palmodictyon*

Class: Chlorophyceae
Order: Sphaeropleales
Family: Scenedesmaceae

Chlorophyta

Comasiella (formerly *Scenedesmus*)

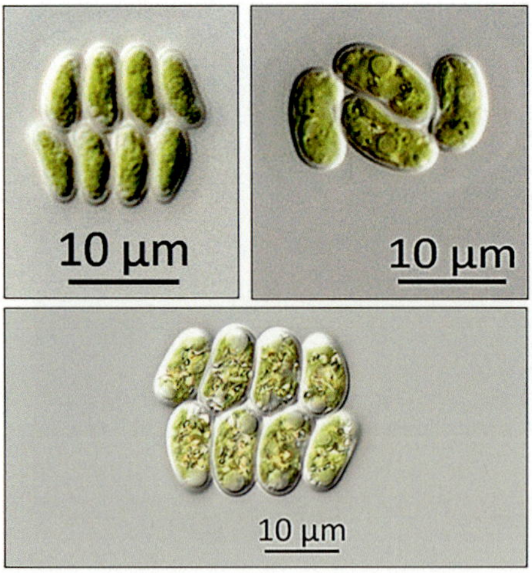

Comasiella arcuata Metaphyton, plankton

Fig. 2.45 *Comasiella* (formerly *Scenedesmus*)

Class: Chlorophyceae
Order: Sphaeropleales
Family: Scenedesmaceae

Chlorophyta

Scenedesmus

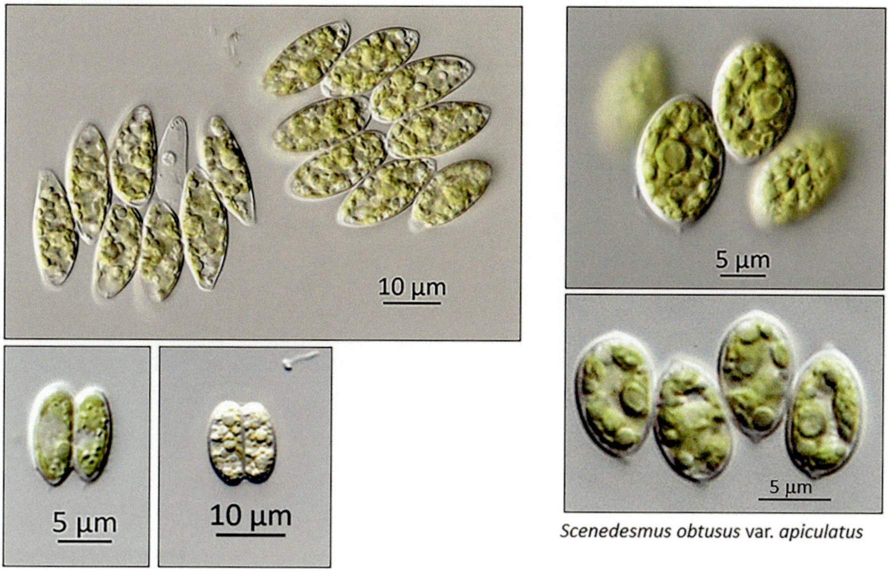

Scenedesmus obtusus var. *apiculatus*

Scenedesmus spp.

Fig. 2.46 *Scenedesmus*

Class: Chlorophyceae
Order: Sphaeropleales
Family: Scenedesmaceae

Chlorophyta

Tetradesmus (formerly *Scenedesmus*)

Tetradesmus dimorphus

Tetradesmus wisconsinensis

Tetradesmus obliquus

Metaphyton, plankton

Fig. 2.47 *Tetradesmus* (formerly *Scenedesmus*)

Class: Chlorophyceae
Order: Sphaeropleales
Family: Scenedesmaceae

Chlorophyta

Pectinodesmus (formerly *Scenedesmus*)

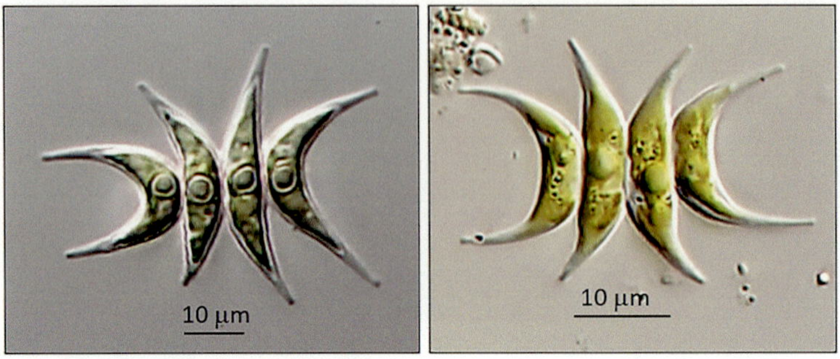

Pectinodesmus pectinatus Metaphyton, plankton

Fig. 2.48 *Pectinodesmus* (formerly *Scenedesmus*)

Class: Chlorophyceae
Order: Sphaeropleales
Family: Scenedesmaceae

Chlorophyta

Desmodesmus

Desmodesmus subspicatus *Desmodesmus bicaudatus* *Desmodesmus* sp.

Desmodesmus serratus Metaphyton, plankton

Fig. 2.49 *Desmodesmus*

Class: Chlorophyceae
Order: Sphaeropleales
Family: Scenedesmaceae

Chlorophyta

Desmodesmus

Desmodesmus perforatus

Desmodesmus communis

Desmodesmus spp.

Metaphyton, plankton

Fig. 2.50 *Desmodesmus*

Class: Chlorophyceae
Order: Sphaeropleales
Family: Scenedesmaceae

Chlorophyta

Tetrastrum

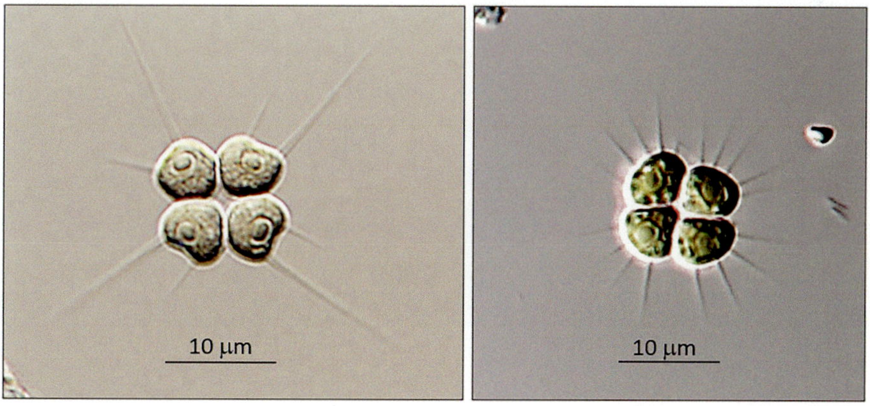

Tetrastrum heteracanthum

Tetrastrum staurogeniiforme

Metaphyton, plankton

Fig. 2.51 *Tetrastrum*

Class: Chlorophyceae
Order: Sphaeropleales
Family: Scenedesmaceae

Chlorophyta

Dimorphococcus

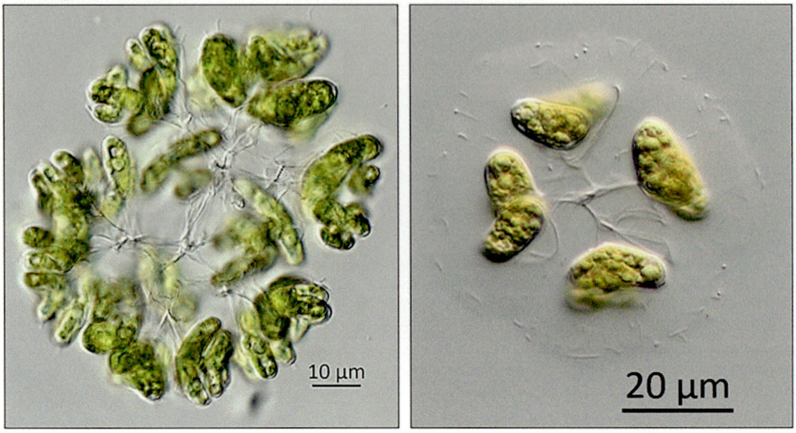

Dimorphococcus lunatus Metaphyton, plankton

Fig. 2.52 *Dimorphococcus*

Class: Chlorophyceae
Order: Sphaeropleales
Family: Scenedesmaceae

Chlorophyta

Coelastrum

Coelastrum cambricum Metaphyton, plankton

Fig. 2.53 *Coelastrum*

Class: Chlorophyceae
Order: Sphaeropleales
Family: Scenedesmaceae

Chlorophyta

Coelastrum

Coelastrum astroideum

Metaphyton, plankton

Fig. 2.54 *Coelastrum*

Class: Chlorophyceae
Order: Sphaeropleales
Family: Scenedesmaceae

Chlorophyta

Coelastrum

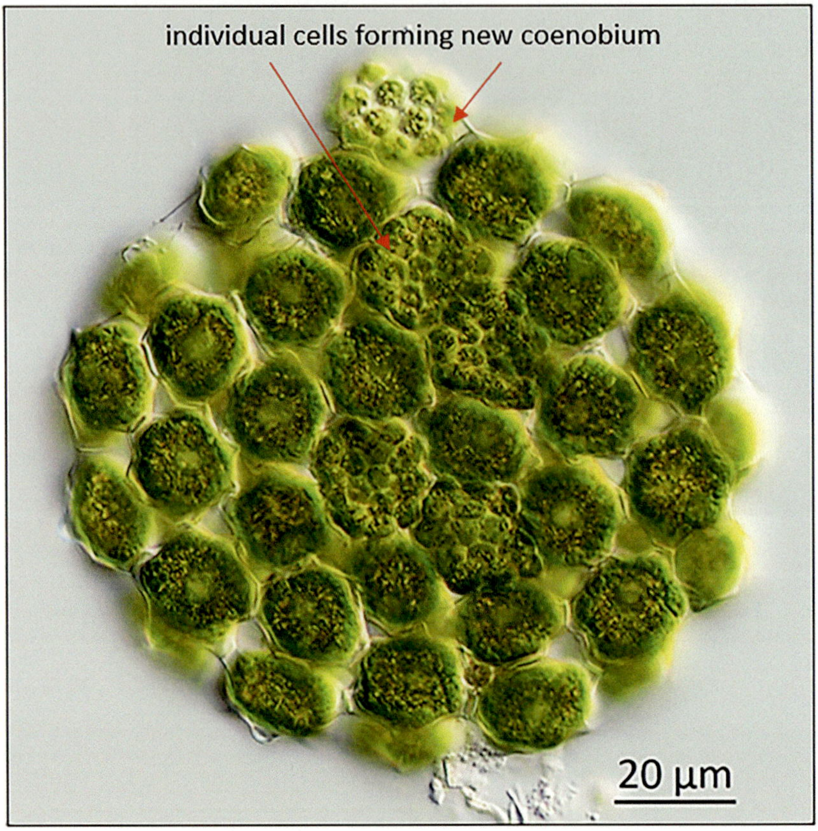

Coelastrum pulchrum

Fig. 2.55 *Coelastrum*

Class: Chlorophyceae
Order: Sphaeropleales
Family: Scenedesmaceae

Chlorophyta

Coelastrum

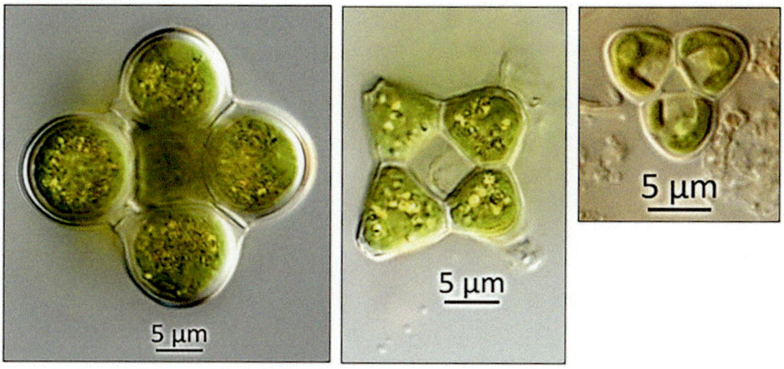

Coelastrum spp.

Metaphyton, plankton

Fig. 2.56 *Coelastrum*

Class: Chlorophyceae
Order: Sphaeropleales
Family: Cylindrocapsaceae

Chlorophyta

Fusola

Fusola viridis

Metaphyton, epiphyton

Fig. 2.57 *Fusola*

Class: Chlorophyceae
Order: Sphaeropleales
Family: Hydrodictyaceae

Chlorophyta

Sorastrum

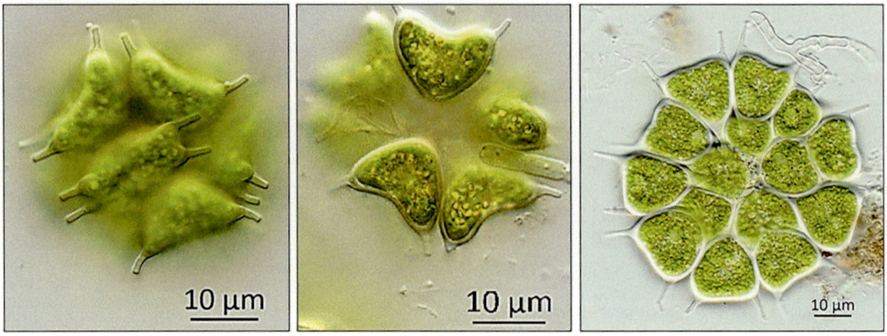

Sorastrum spinulosum Metaphyton, plankton

Fig. 2.58 *Sorastrum*

Class: Chlorophyceae
Order: Sphaeropleales
Family: Hydrodictyaceae

Chlorophyta

Sorastrum

Sorastrum americanum Metaphyton

Fig. 2.59 *Sorastrum*

Class: Chlorophyceae
Order: Chaetophorales
Family: Aphanochaetaceae

Chlorophyta

Aphanochaete

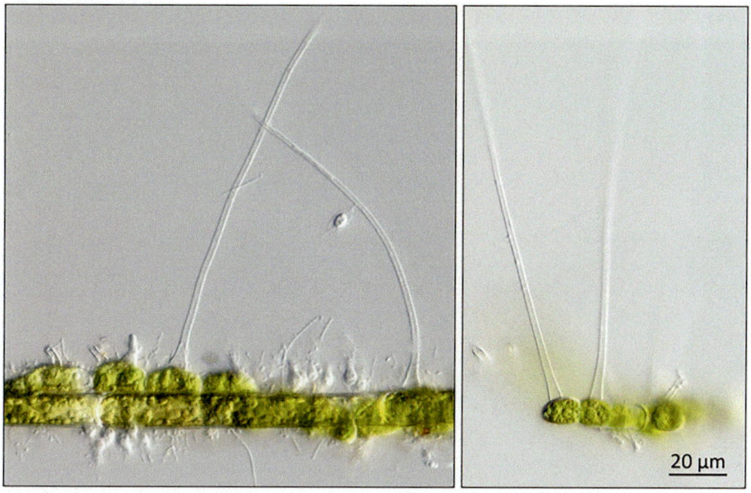

Aphanochaete polychaete

Epiphyton

Fig. 2.60 *Aphanochaete*

Class: Chlorophyceae
Order: Sphaeropleales
Family: Hydrodictyceae

Chlorophyta

Hydrodictyon

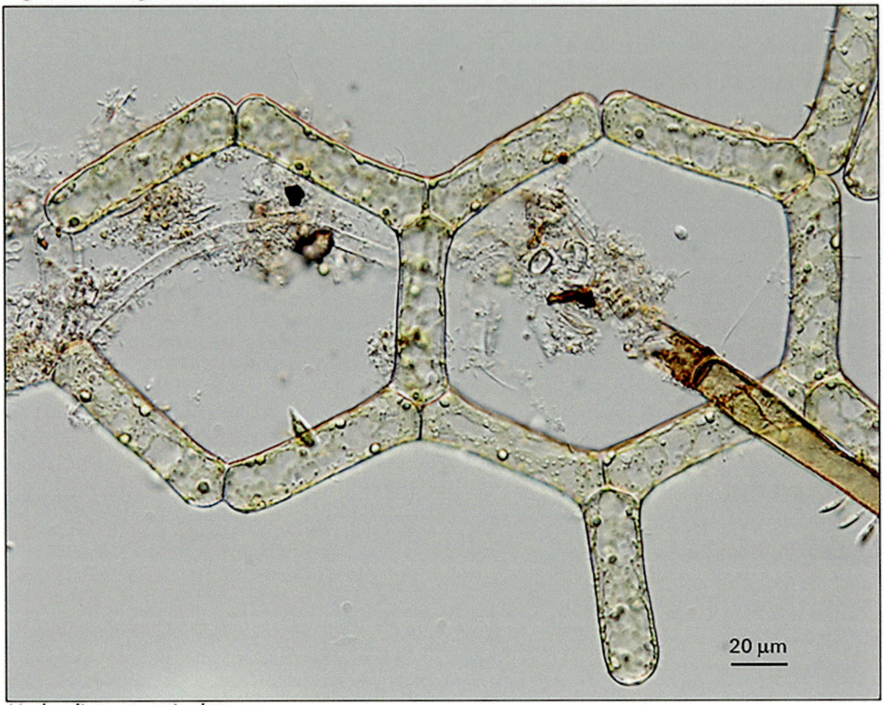

Hydrodictyon reticulatum

Fig. 2.61 *Hydrodictyon*

Class: Chlorophyceae
Order: Sphaeropleales
Family: Hydrodictyaceae

Chlorophyta

Tetraedron

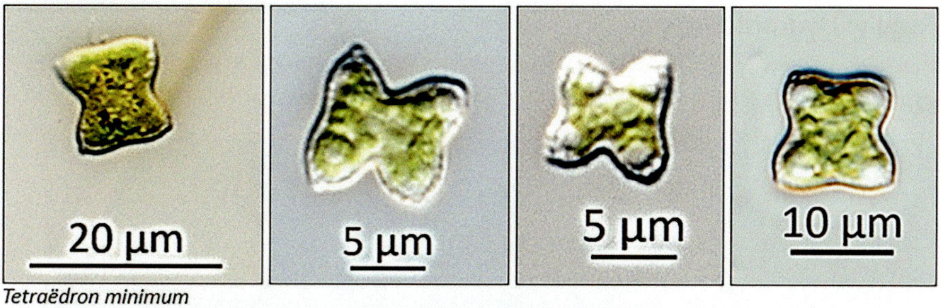

Tetraëdron minimum

Tetraëdron caudatum

Tetraëdron trigonum

Fig. 2.62 *Tetraedron*

Class: Chlorophyceae
Order: Sphaeropleales
Family: Hydrodictyaceae

Chlorophyta

Pediastrum

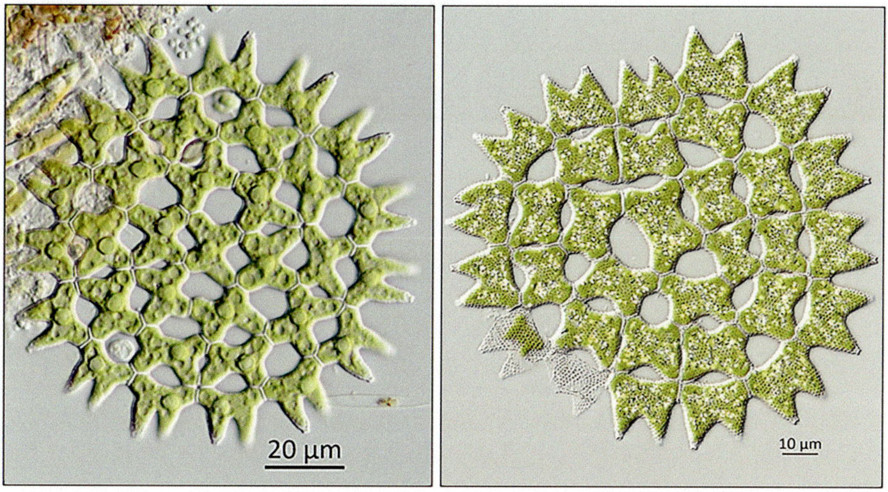

Pediastrum duplex

Fig. 2.63 *Pediastrum*

Class: Chlorophyceae
Order: Sphaeropleales
Family: Hydrodictyaceae

Chlorophyta

Pediastrum

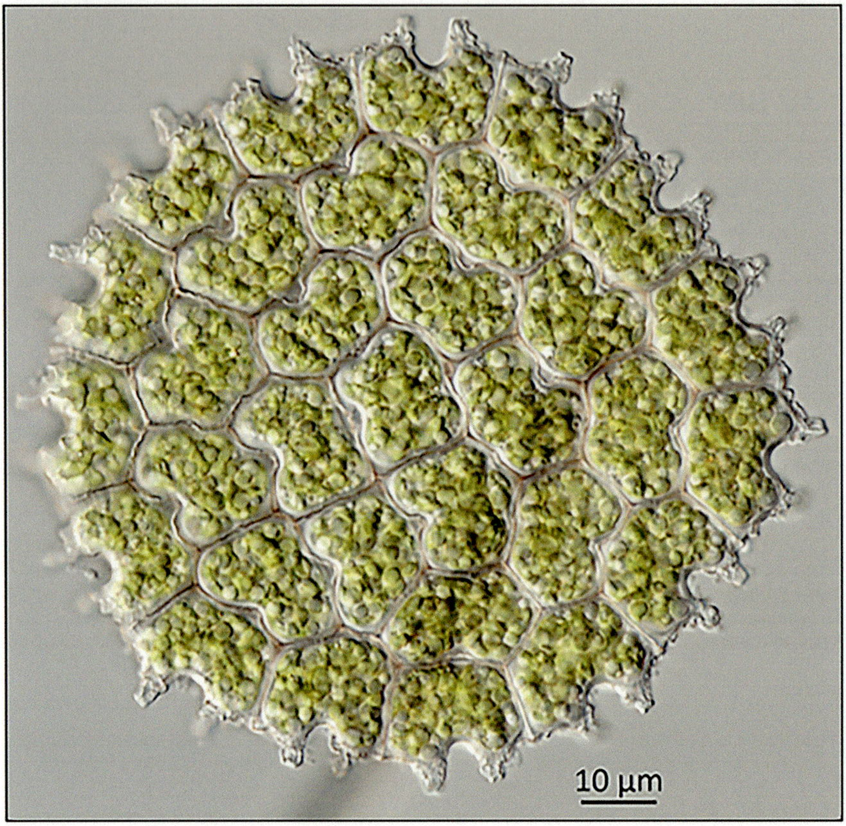

Pediastrum angulosum

Fig. 2.64 *Pediastrum*

Class: Chlorophyceae
Order: Sphaeropleales
Family: Hydrodictyaceae

Chlorophyta

Pseudopediastrum (formerly *Pediastrum*)

Pseudopediastrum boryanum

Fig. 2.65 *Pseudopediastrum* (formerly *Pediastrum*)

Class: Chlorophyceae
Order: Sphaeropleales
Family: Hydrodictyaceae

Chlorophyta

Monactinus (formerly *Pediastrum*)

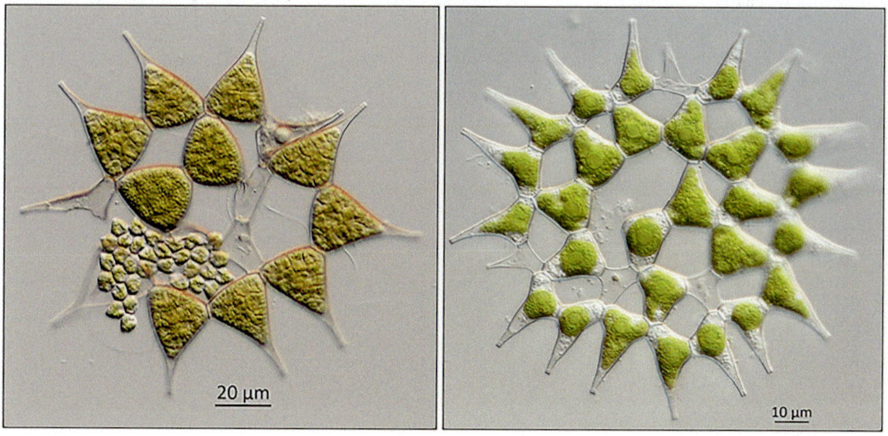

Monactinus simplex

Fig. 2.66 *Monactinus (formerly Pediastrum)*

Class: Chlorophyceae
Order: Sphaeropleales
Family: Hydrodictyaceae

Chlorophyta

Monactinus (formerly *Pediastrum*)

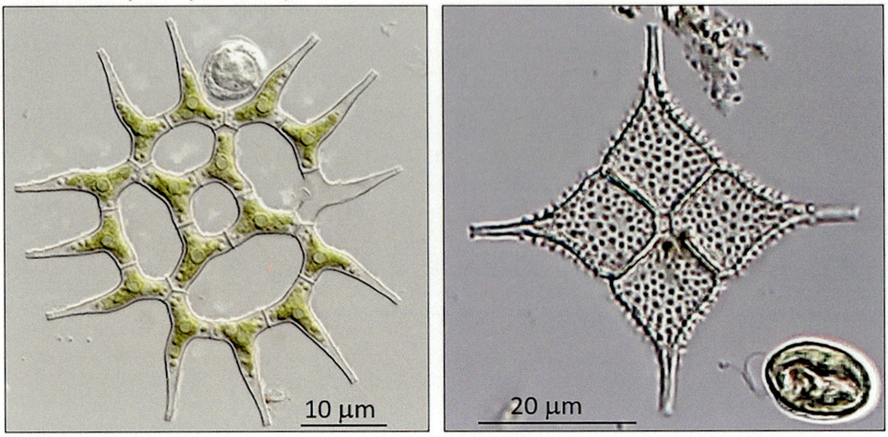

Monactinus simplex *Monactinus simplex* var. *echinulatum*

Fig. 2.67 *Monactinus (formerly Pediastrum)*

Class: Chlorophyceae
Order: Sphaeropleales
Family: Hydrodictyaceae

Chlorophyta

Lacunastrum (formerly *Pediastrum*)

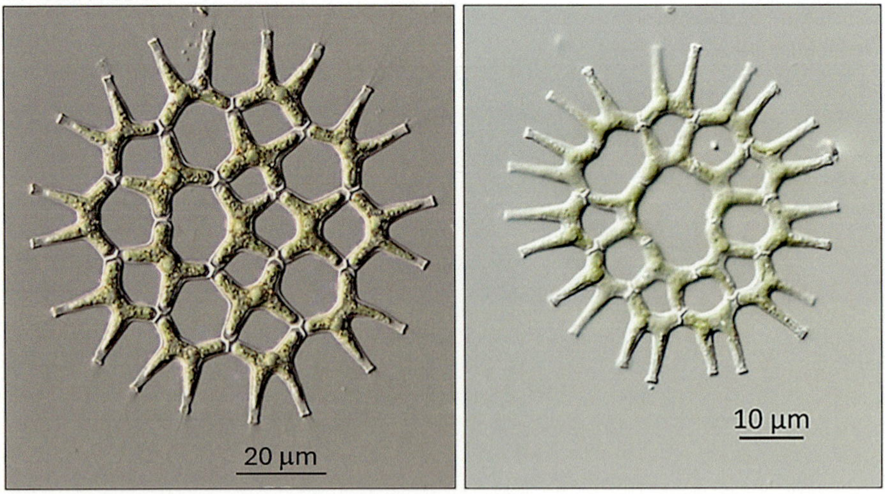

Lacunastrum gracillimum

Fig. 2.68 *Lacunastrum (formerly Pediastrum)*

Class: Chlorophyceae
Order: Sphaeropleales
Family: Hydrodictyaceae

Chlorophyta

Stauridium (formerly *Pediastrum*)

Stauridium tetras

Fig. 2.69 *Stauridium (formerly Pediastrum)*

Class: Chlorophyceae
Order: Sphaeropleales
Family: Schroederiaceae

Chlorophyta

Schroederia

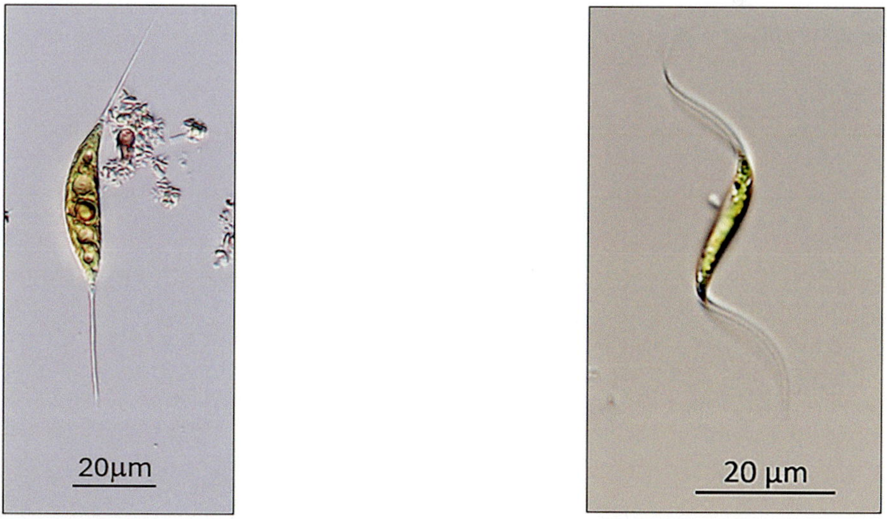

Schroederia setigera

Schroederia spiralis

Fig. 2.70 *Schroederia*

Class: Chlorophyceae
Order: Sphaeropleales
Family: Schroederiaceae

Chlorophyta

Ankyra

Ankyra judayi

Fig. 2.71 *Ankyra*

Class: Chlorophyceae
Order: Sphaeropleales
Family: Neochloridaceae

Chlorophyta

Golenkinia

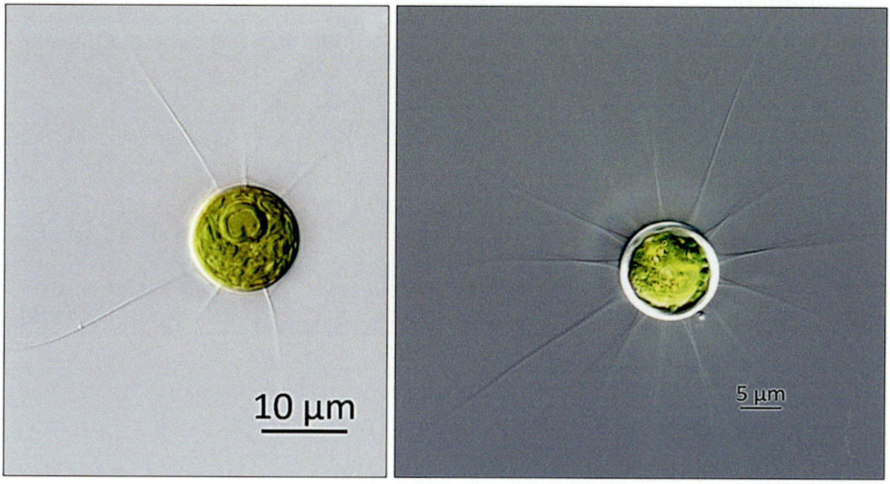

Golenkinia radiata

Fig. 2.72 *Golenkinia*

Class: Chlorophyceae
Order: Sphaeropleales
Family: Treubariaceae

Chlorophyta

Treubaria

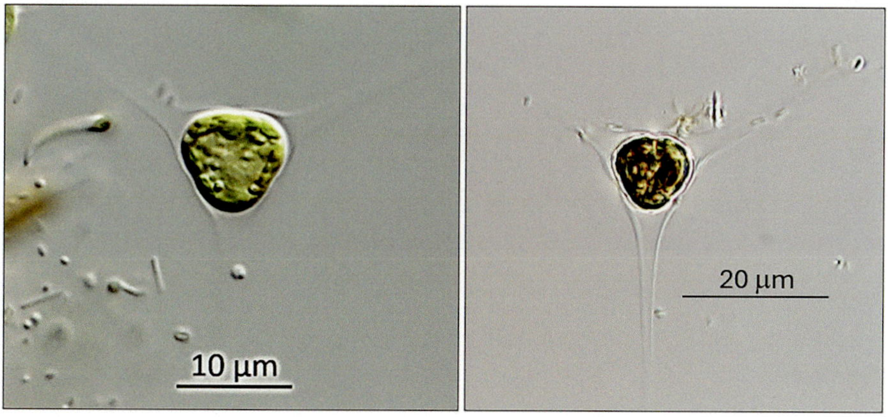

Treubaria schmidlei

Fig. 2.73 *Treubaria*

Class: Chlorophyceae
Order: Chaetopeltidales
Family: Hydrocytiaceae

Chlorophyta

Hydrocytium

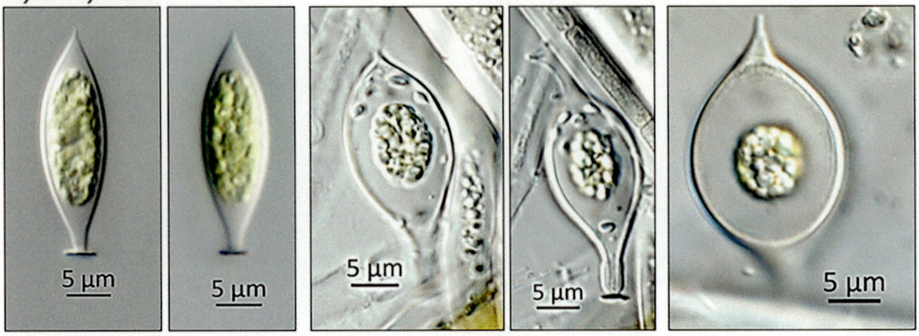

Hydrocytium acuminatum

Fig. 2.74 *Hydrocytium*

Class: Trebouxiophyceae
Order: Chlorellales
Family: Chlorellaceae

Chlorophyta

Chlorella

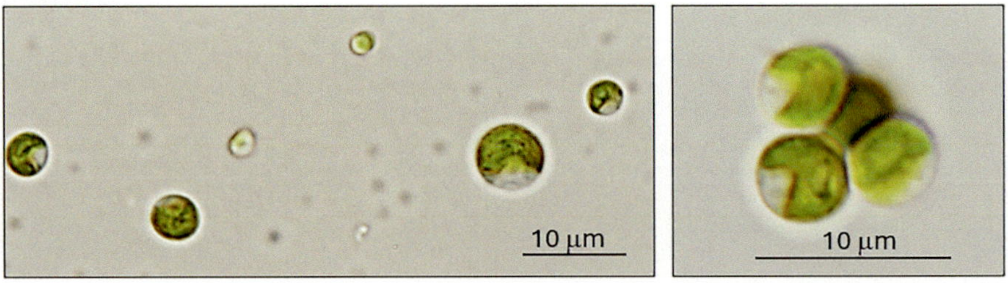

Chlorella sp.

Fig. 2.75 *Chlorella*

Class: Trebouxiophyceae
Order: Chlorellales
Family: Chlorellaceae

Chlorophyta

Actinastrum

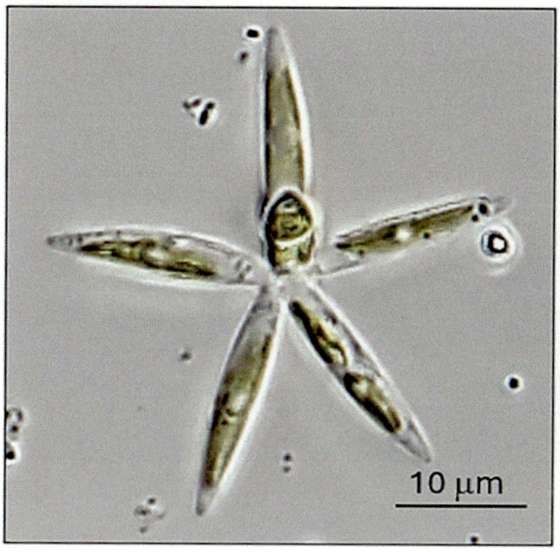

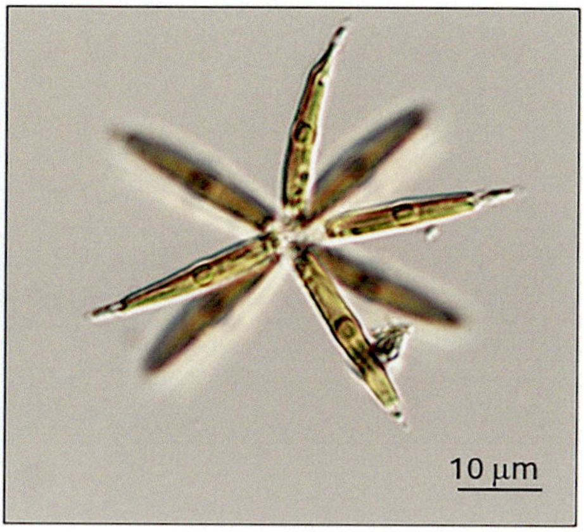

Actinastrum fluviatile

Fig. 2.76 *Actinastrum*

Class: Trebouxiophyceae
Order: Chlorellales
Family: Chlorellaceae

Chlorophyta

Closteriopsis

Closteriopsis brevicula

Fig. 2.77 *Closteriopsis*

Class: Trebouxiophyceae
Order: Chlorellales
Family: Chlorellaceae

Chlorophyta

Geminella

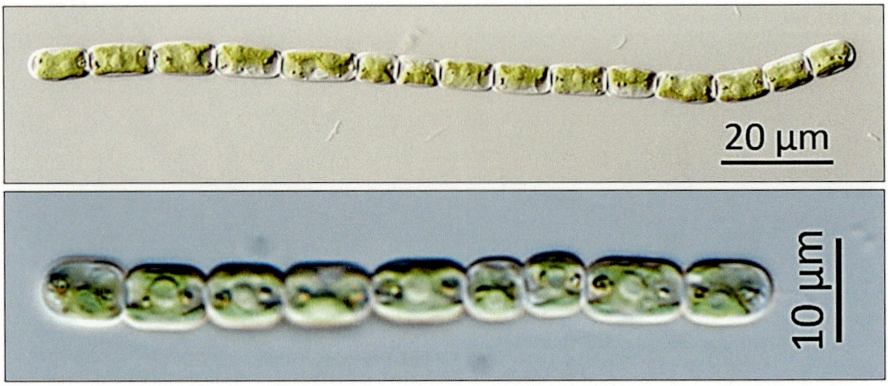

Geminella mutabilis (Brébisson) Wille 1909: 72.

Fig. 2.78 *Geminella*

Class: Trebouxiophyceae
Order: Chlorellales
Family: Oocystaceae

Chlorophyta

Gloeotaenium

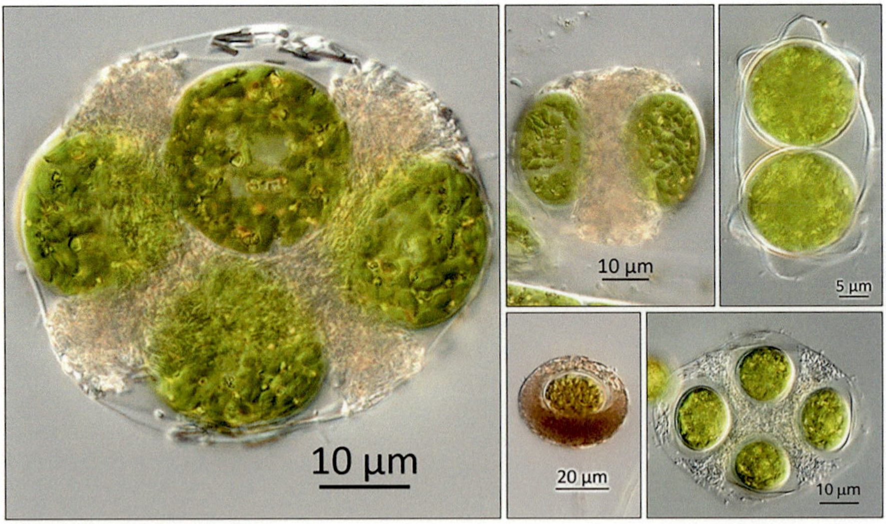

Gloeotaenium loitlesbergerianum

Fig. 2.79 *Gloeotaenium*

Class: Trebouxiophyceae
Order: Chlorellales
Family: Oocystaceae

Chlorophyta

Oocystis

Oocystis spp.

Fig. 2.80 *Oocystis*

Class: Trebouxiophyceae
Order: Chlorellales
Family: Oocystaceae

Chlorophyta

Franceia

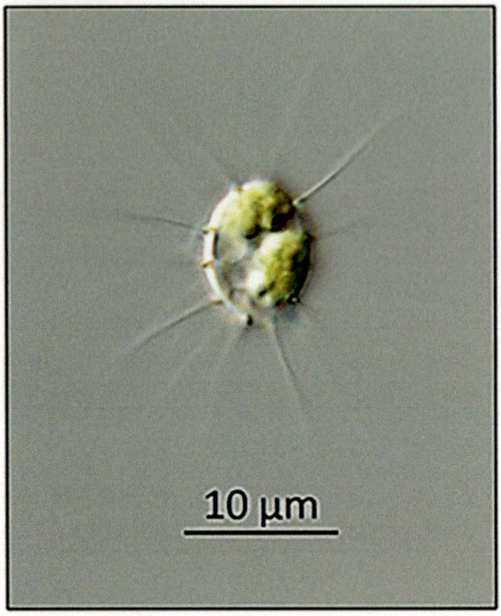

Franceia ovalis (Francé 1894) Lemmermann 1898

Fig. 2.81 *Franceia*

Class: Trebouxiophyceae
Order: Chlorellales
Family: Oocystaceae

Chlorophyta

Willea

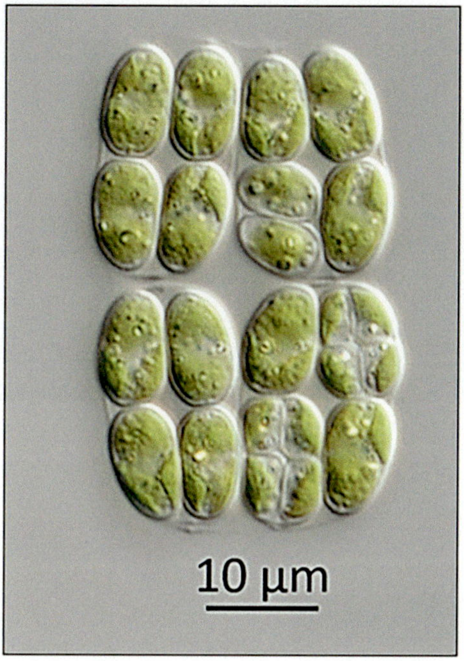

Willea apiculata

Fig. 2.82 *Willea*

Class: Trebouxiophyceae
Order: Chlorellales
Family: Chlorellaceae

Chlorophyta

Micractinium

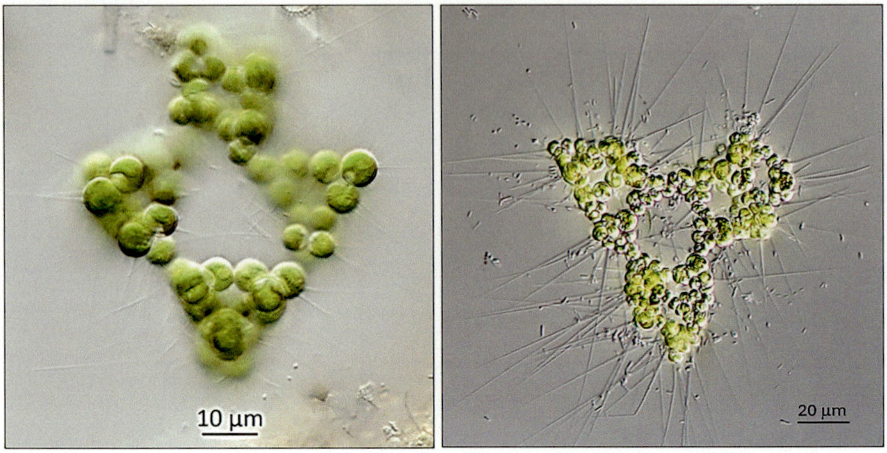

Micractinium bornhemiense

Fig. 2.83 *Micractinium*

Class: Trebouxiophyceae
Order: Chlorellales
Family: Chlorellaceae

Chlorophyta

Mucidosphaerium (formerly *Dictyosphaerium*)

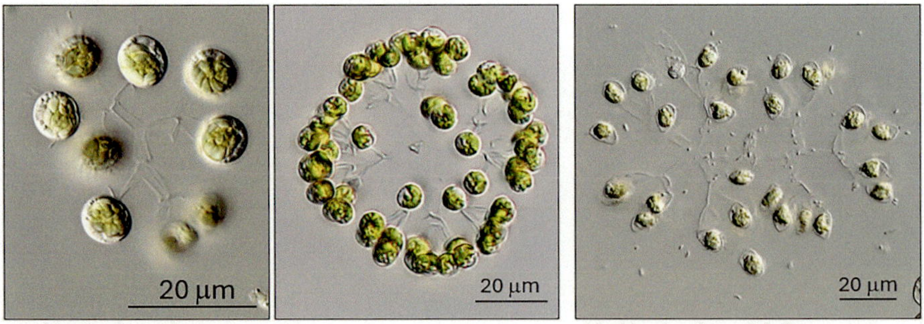

Mucidosphaerium palustre *Mucidosphaerium pulchellum*

Fig. 2.84 *Mucidosphaerium* (formerly *Dictyosphaerium*)

Class: Trebouxiophyceae
Order: Chlorellales
Family: Oocystaceae

Chlorophyta

Lagerheimia

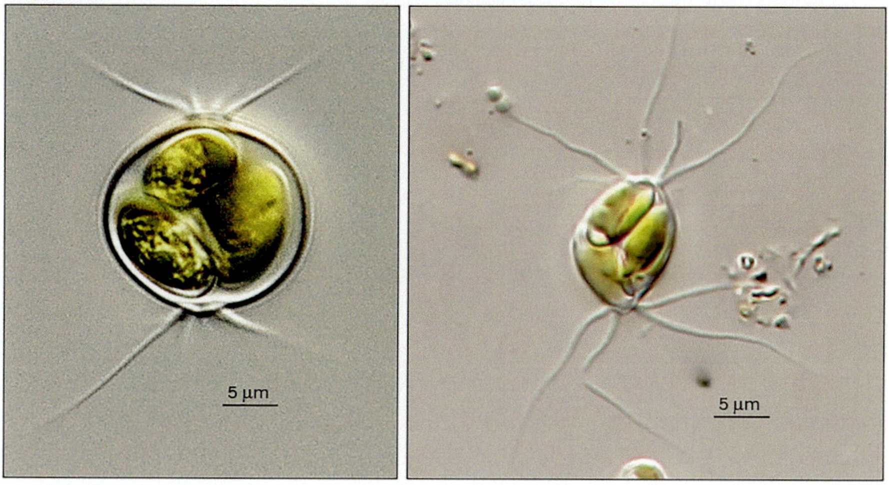

Lagerheimia ciliata

Fig. 2.85 *Lagerheimia*

Class: Trebouxiophyceae
Order: Chlorellales
Family: Nephrocytiaceae

Chlorophyta

Nephrocytium

Nephrocytium lunatum

Nephrocytium agardhianum

Nephrocytium limneticum

Fig. 2.86 *Nephrocytium*

Class: Trebouxiophyceae
Order: Trebouxiales
Family: Botryococcaceae

Chlorophyta

Botryococcus

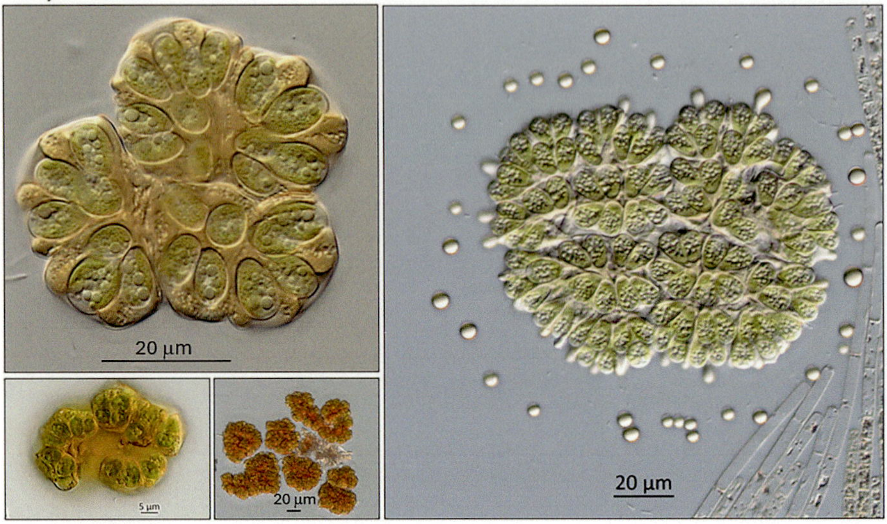

Botryococcus braunii

Fig. 2.87 *Botryococcus*

Class: Trebouxiophyceae
Order: Trebouxiophyceae *ordo incertae sedis*
Family: Coccomyxaceae

Chlorophyta

Dispora

Dispora crucigenioides

Fig. 2.88 *Dispora*

Class: Trebouxiophyceae
Order: Trebouxiophyceae *ordo incertae sedis*
Family: Coccomyxaceae

Chlorophyta

Choricystis

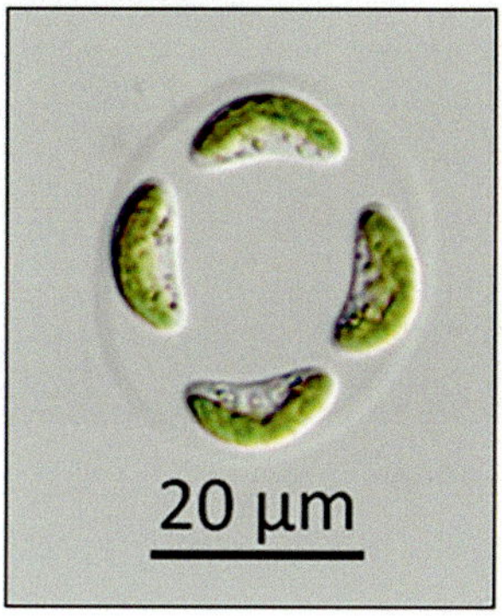

Choricystis nana

Fig. 2.89 *Choricystis*

Class: Trebouxiophyceae
Order: Trebouxiophyceae *ordo incertae sedis*
Family: Trebouxiophyceae *incertae sedis*

Chlorophyta

Lemmermannia

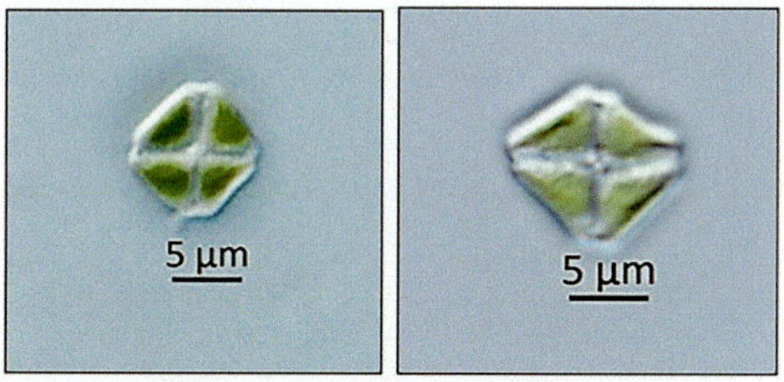

Note: both the Order and Family of this genus is uncertain

Lemmermannia tetrapedia

Phytoplankton

Fig. 2.90 *Lemmermannia*

Class: Nephroselmidophyeae
Order: Nephroselmidales
Family: Nephroselmidaceae

Chlorophyta

Nephroselmis

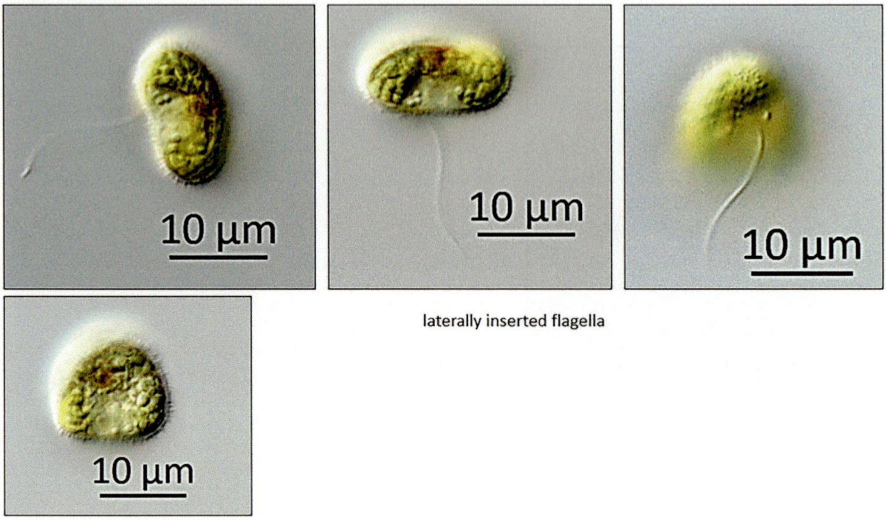

laterally inserted flagella

Nephroselmis olivacea

Fig. 2.91 *Nephroselmis*

Class: Pyraminomadophyceae
Order: Pyramimonadales
Family: Pyramimonadaceae

Chlorophyta

Pyramimonas

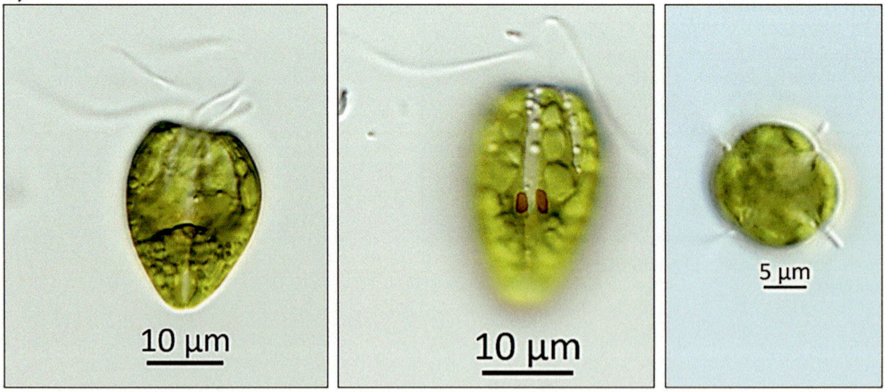

Pyramimonas sp.

Plankton

Fig. 2.92 *Pyramimonas*

Class: Ulvophyceae
Order: Cladophorales
Family: Cladophoraceae

Chlorophyta

Rhizoclonium

Rhizoclonium spp. Kützing, F.T. (1843)

Fig. 2.93 *Rhizoclonium*

Class: Ulvophyceae
Order: Ulotrichales
Family: Ulotrichaceae

Chlorophyta

Ulothrix

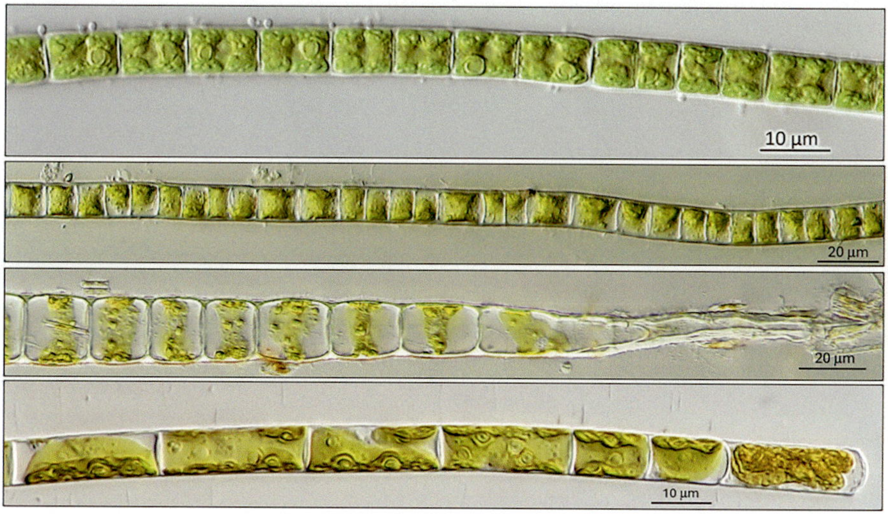

Ulothrix sp.

Fig. 2.94 *Ulothrix*

Class: Ulvophyceae
Order: Cladophorales
Family: Cladophoraceae

Chlorophyta

Cladophora

Cladophora spp.

Fig. 2.95 *Cladophora*

Class: Ulvophyceae
Order: Bryopsidales
Family: Dichotomosiphonaceae

Chlorophyta

Dichotomosiphon

Dichotomosiphon tuberosus

Fig. 2.96 *Dichotomosiphon*

Class: Ulvophyceae
Order: Cladophorales
Family: Pithophoraceae

Chlorophyta

Pithophora

Pithophora roettleri

Fig. 2.97 *Pithophora*

Class: Ulvophyceae
Order: Cladophorales
Family: Pithophoraceae

Chlorophyta

Arnoldiella (formerly *Basicladia*)

Arnoldiella crassa

Epizoic on turtles and alligators

Fig. 2.98 *Arnoldiella* (formerly *Basicladia*)

Class: Ulvophyceae
Order: Cladophorales
Family: Pithophoraceae

Chlorophyta

Arnoldiella (formerly *Basicladia*)

prostrate layer consisting of coalescing filaments

upright filaments

Arnoldiella crassa

Epizoic on turtles and alligators

Fig. 2.99 *Arnoldiella* (formerly *Basicladia*)

Class: Ulvophyceae
Order: Ulotrichales
Family: Helicodictyaceae

Chlorophyta

Protoderma (cf)

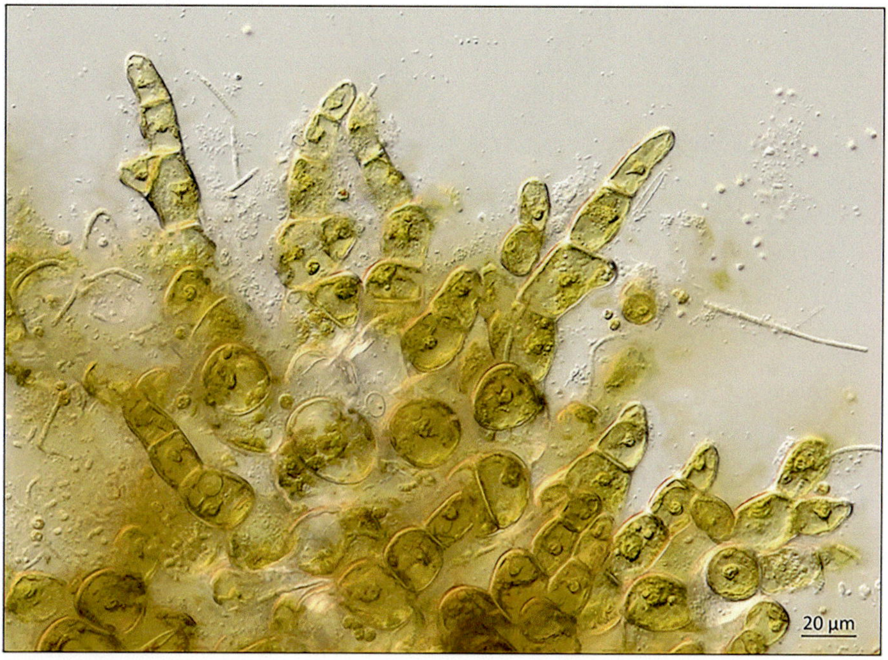

Protoderma sp.

Fig. 2.100 *Protoderma* (cf)

Class: Ulvophyceae
Order: Ulvales
Family: Ulvaceae

Chlorophyta

Ulva (formerly *Enteromorpha*)

Ulva sp. Periphyton

Fig. 2.101 *Ulva* (formerly *Enteromorpha*)

Class: Ulvophyceae
Order: Ulotrichales
Family: Monostromataceae

Chlorophyta

Monostroma

Monostroma bullosum

Fig. 2.102 *Monostroma*

Class: Charophyceae
Order: Charales
Family: Characeae

Charophyta

Chara

Chara sp.

"rooted" in benthic substratum

Fig. 2.103 *Chara*

Class: Coleochaetophyceae
Order: Chaetosphaeridiales
Family: Chaetosphaeridiaceae

Charophyta

Chaetosphaeridium

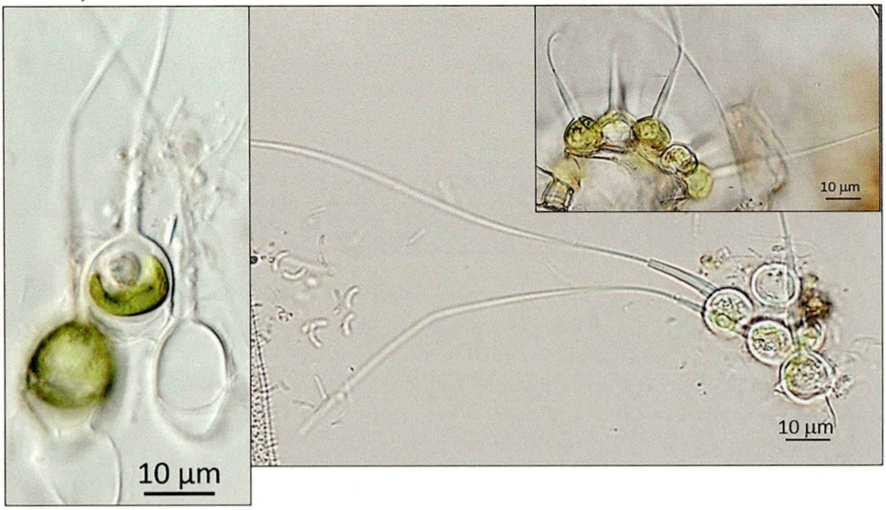

Chaetosphaeridium spp.

Epiphyton

Fig. 2.104 *Chaetosphaeridium*

Class: Chlorokybophyceae
Order: Chlorokybales
Family: Chlorokybaceae

Charophyta

Chlorokybus

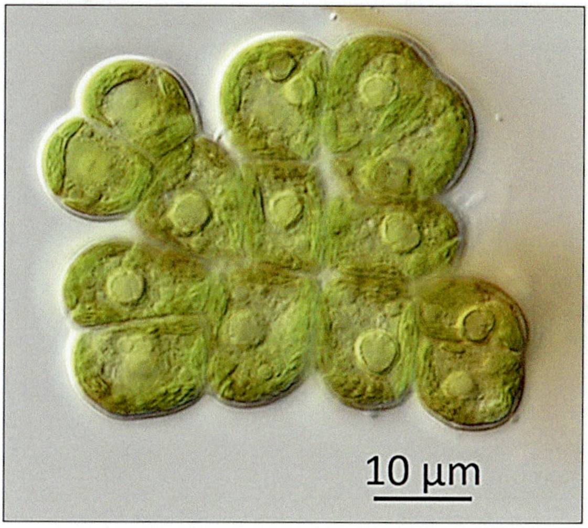

Chlorokybus atmophyticus Epiphyton, epilithon

Fig. 2.105 *Chlorokybus*

Class: Klebsormidiophyceae
Order: Klebsormidiales
Family: Elkatotrichaceae

Charophyta

Elakatothrix

Elakatothrix gelatinosa Metaphyton

Fig. 2.106 *Elakatothrix*

Class: Zygnemataphyceae
Order: Desmidiales
Family: Closteriaceae

Charophyta

Closterium

Closterium spp.

Metaphyton

Fig. 2.107 *Closterium*

Class: Zygnemataphyceae
Order: Desmidiales
Family: Closteriaceae

Charophyta

Closterium

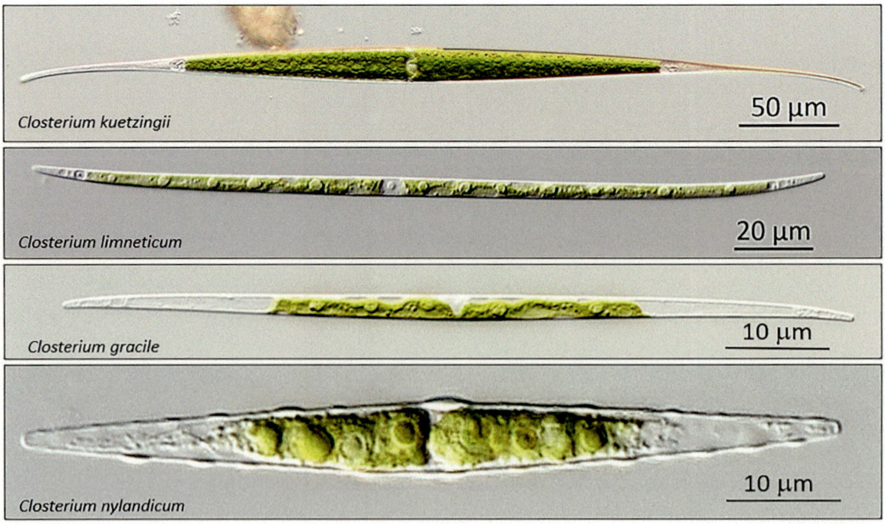

Metaphyton

Fig. 2.108 *Closterium*

Class: Zygnematophyceae
Order: Desmidiales
Family: Closteriaceae

Charophyta

Spinoclosterium

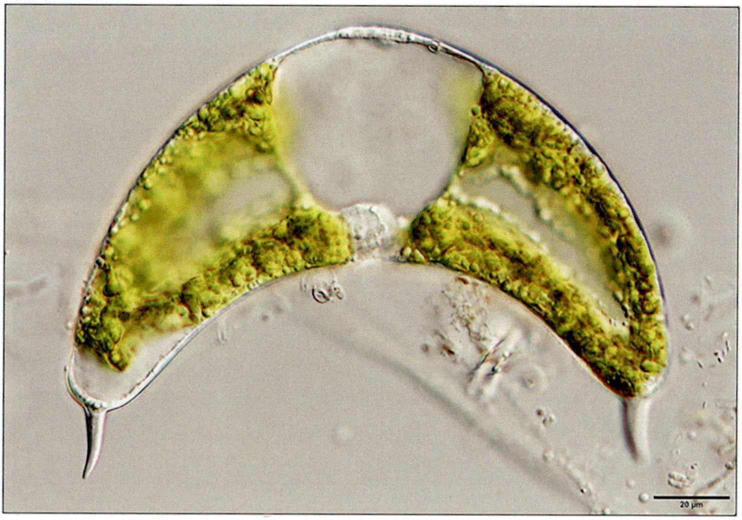

Spinoclosterium curvatum Metaphyton

Fig. 2.109 *Spinoclosterium*

Class: Zygnemataphyceae
Order: Desmidiales
Family: Desmidiaceae

Charophyta

Actinotaenium

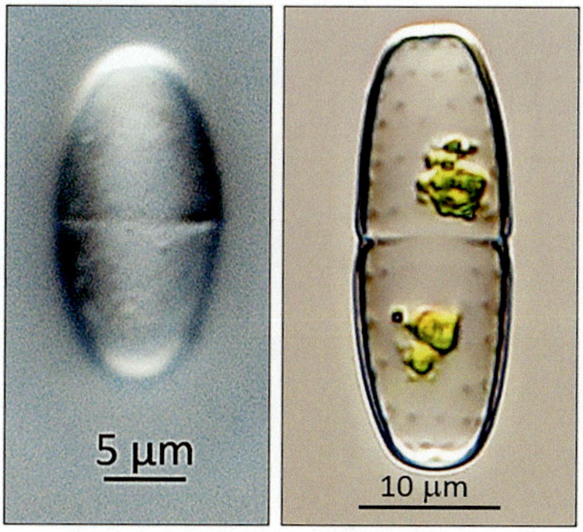

Actinotaenium spp. Metaphyton

Fig. 2.110 *Actinotaenium*

Class: Zygnemataphyceae
Order: Desmidiales
Family: Desmidiaceae

Charophyta

Cosmarium

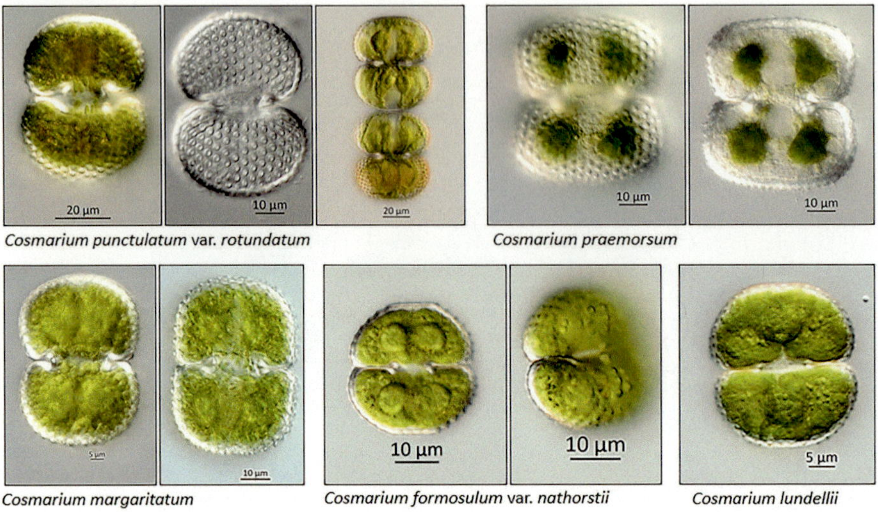

Metaphyton

Fig. 2.111 *Cosmarium*

Class: Zygnemataphyceae
Order: Desmidiales
Family: Desmidiaceae

Charophyta

Cosmarium

Cosmarium contractum var. *ellipsoideum*
Cosmarium galeritum

Metaphyton

Fig. 2.112 *Cosmarium*

Class: Zygnemataphyceae
Order: Desmidiales
Family: Desmidiaceae

Charophyta

Cosmarium

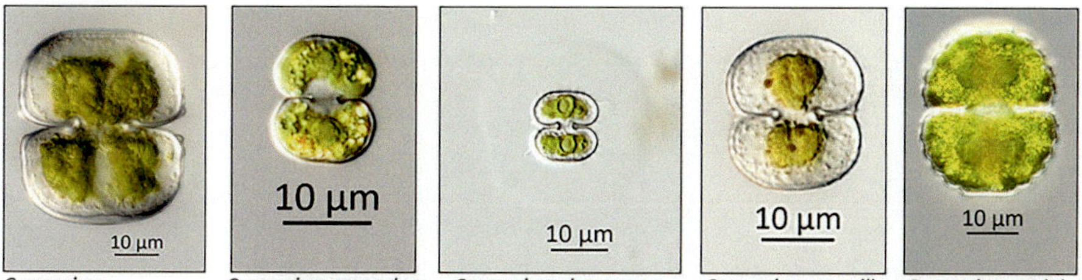

Cosmarium smolandicum | *Cosmarium repandum f. sexangulare* | *Cosmarium depressum* var. *reniforme* | *Cosmarium mamill iferum* | *Cosmarium undula tum* var. *minutum*

Cosmarium subtumidum | *Cosmarium tenue* | *Cosmarium angulosum* | *Cosmarium didymoprotupsum*
Metaphyton

Fig. 2.113 *Cosmarium*

Class: Zygnemataphyceae
Order: Desmidiales
Family: Desmidiaceae

Charophyta

Cosmarium

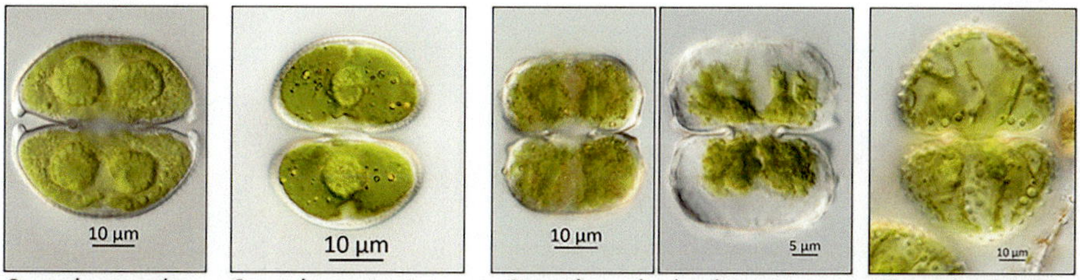

Cosmarium pseudo-axichondrum var. *foggii* | *Cosmarium contractum* var. *pachydermum* | *Cosmarium anisochondrum* var. *geminatum* | *Cosmarium botrytis* var. *tumidum*

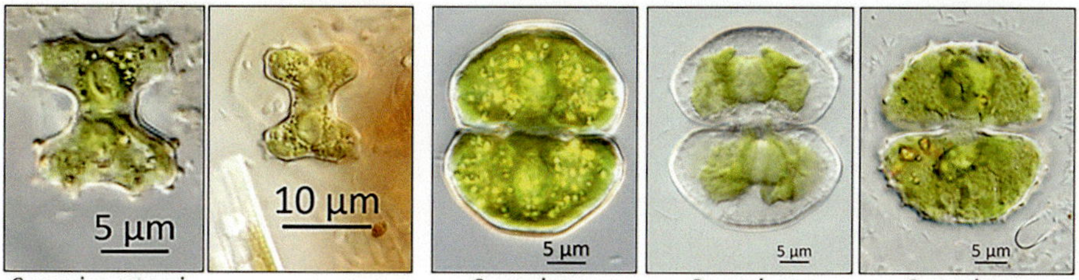

Cosmarium regnesi | *Cosmarium* sp. | *Cosmarium* sp. | *Cosmarium* sp.
Metaphyton

Fig. 2.114 *Cosmarium*

Class: Zygnemataphyceae
Order: Desmidiales
Family: Desmidiaceae

Charophyta

Desmidium

Metaphyton

Fig. 2.115 *Desmidium*

Class: Zygnemataphyceae
Order: Desmidiales
Family: Desmidiaceae

Charophyta

Euastrum

Euastrum spp.

Metaphyton

Fig. 2.116 *Euastrum*

Class: Zygnemataphyceae
Order: Desmidiales
Family: Desmidiaceae

Charophyta

Euastrum

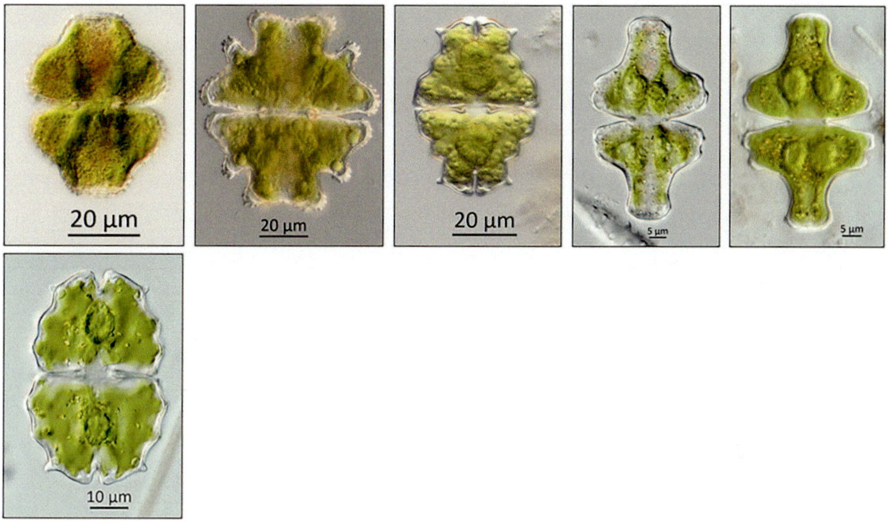

Euastrum spp. Metaphyton

Fig. 2.117 *Euastrum*

Class: Zygnemataphyceae
Order: Desmidiales
Family: Desmidiaceae

Charophyta

Euastrum

Euastrum verrucosum var. *alatum* Metaphyton

Fig. 2.118 *Euastrum*

Class: Zygnemataphyceae
Order: Desmidiales
Family: Desmidiaceae

Charophyta

Gonatozygon

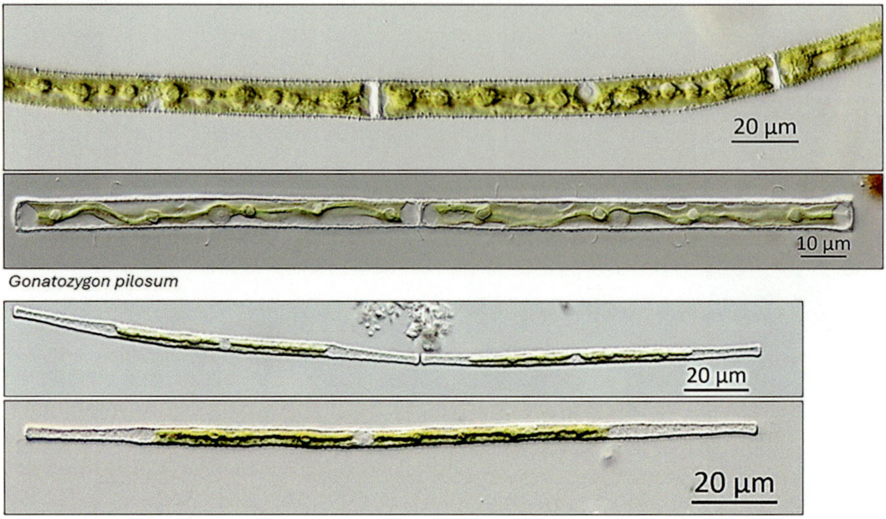

Gonatozygon brebissonii

Metaphyton

Fig. 2.119 *Gonatozygon*

Class: Zygnemataphyceae
Order: Desmidiales
Family: Desmidiaceae

Charophyta

Groenbladia

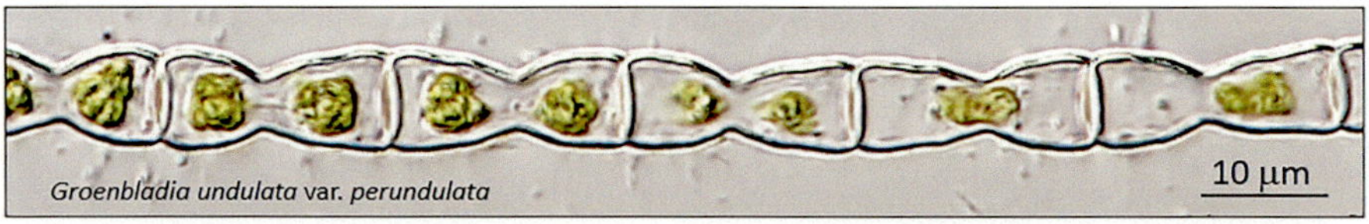

Hyalotheca

Metaphyton

Fig. 2.120 *Groenbladia* and *Hyalotheca*

Class: Zygnemataphyceae
Order: Desmidiales
Family: Desmidiaceae

Charophyta

Haplotaenium

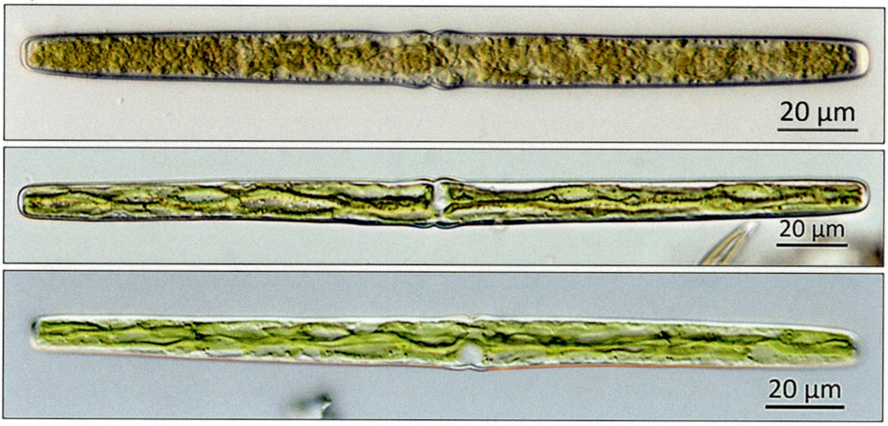

Haplotaenium minutum

Metaphyton

Fig. 2.121 *Haplotaenium*

Class: Zygnemataphyceae
Order: Desmidiales
Family: Desmidiaceae

Charophyta

Heimansia

Heimansia pusilla

Metaphyton

Fig. 2.122 *Heimansia*

Class: Zygnemataphyceae
Order: Desmidiales
Family: Desmidiaceae

Charophyta

Micrasterias

Micrasterias radiosa Metaphyton

Fig. 2.123 *Micrasterias*

Class: Zygnemataphyceae
Order: Desmidiales
Family: Desmidiaceae

Charophyta

Micrasterias

Micrasterias radians Metaphyton

Fig. 2.124 *Micrasterias*

Class: Zygnemataphyceae
Order: Desmidiales
Family: Desmidiaceae

Charophyta

Micrasterias

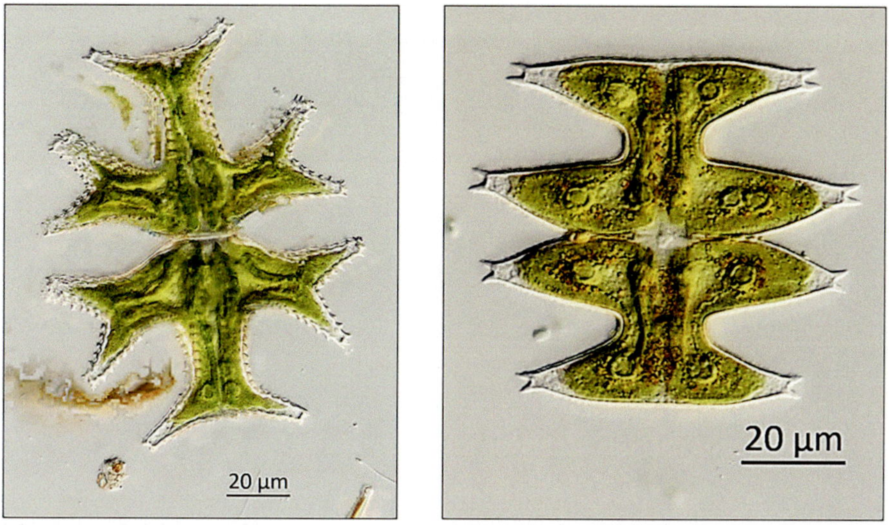

Micrasterias mahabuleshwarensis *Micrasterias pinnatifida* Ralfs 1848 Metaphyton

Fig. 2.125 *Micrasterias*

Class: Zygnemataphyceae
Order: Desmidiales
Family: Desmidiaceae

Charophyta

Micrasterias

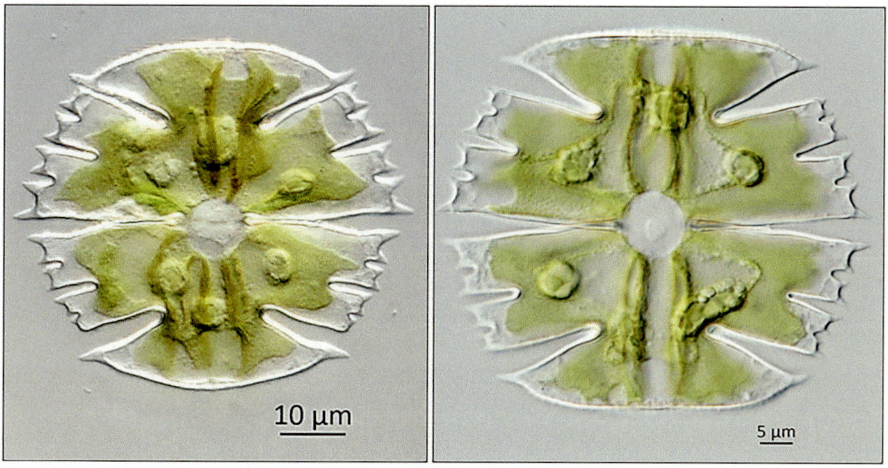

Micrasterias truncata var. *pusilla* Metaphyton

Fig. 2.126 *Micrasterias*

Class: Zygnemataphyceae
Order: Desmidiales
Family: Desmidiaceae

Charophyta

Onychonema

Onychonema filiforme

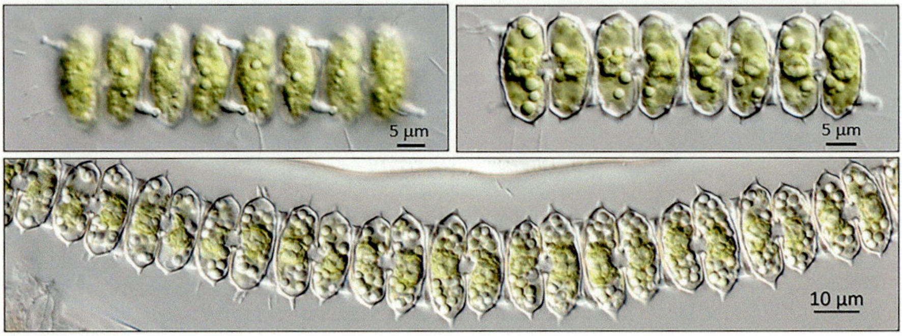

Onychonema sp.

Metaphyton

Fig. 2.127 *Onychonema*

Class: Zygnemataphyceae
Order: Desmidiales
Family: Desmidiaceae

Charophyta

Spondylosium

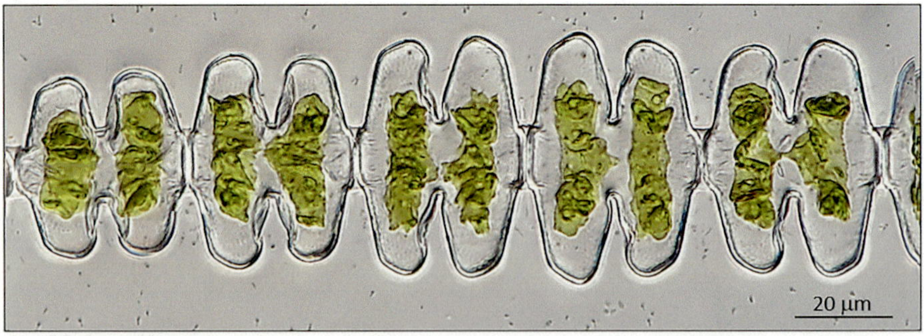

Spondylosium pulchrum

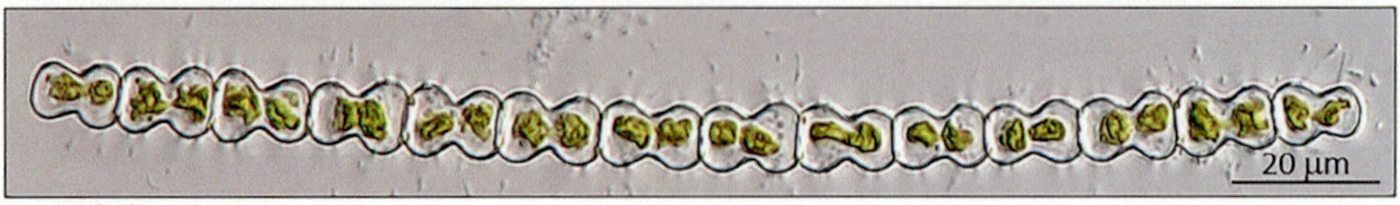

Spondylosium planum

Metaphyton

Fig. 2.128 *Spondylosium*

Class: Zygnemataphyceae
Order: Desmidiales
Family: Desmidiaceae

Charophyta

Pleurotaenium

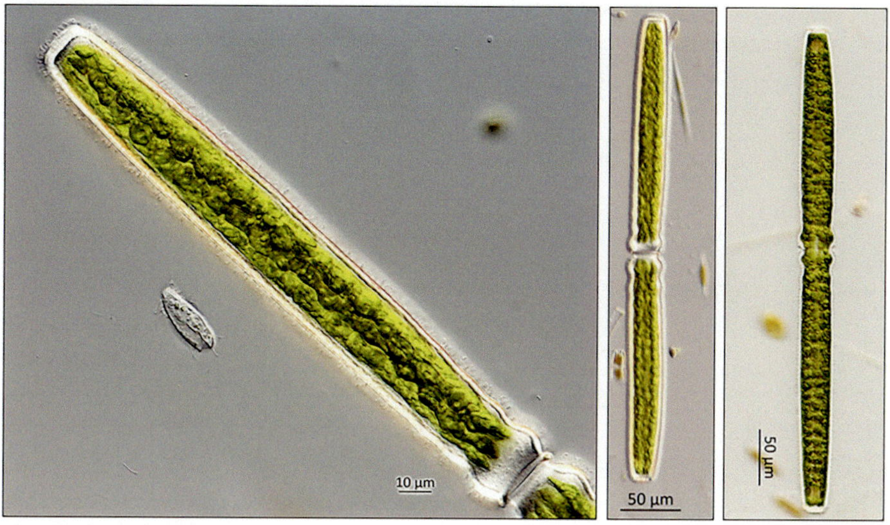

Fig. 2.129 *Pleurotaenium*

Class: Zygnemataphyceae
Order: Desmidiales
Family: Desmidiaceae

Charophyta

Staurastrum

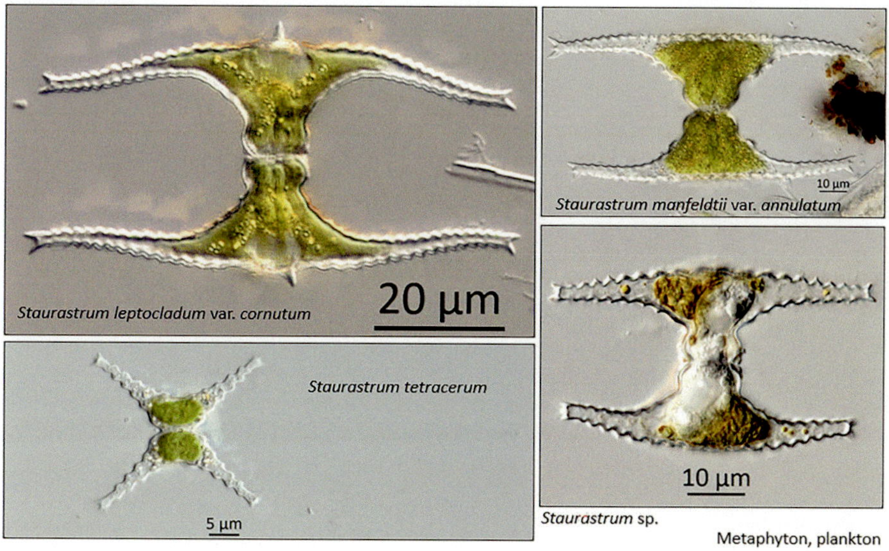

Fig. 2.130 *Staurastrum*

Class: Zygnemataphyceae
Order: Desmidiales
Family: Desmidiaceae

Charophyta

Staurastrum

Staurastrum spp. Metaphyton, plankton

Fig. 2.131 *Staurastrum*

Class: Zygnemataphyceae
Order: Desmidiales
Family: Desmidiaceae

Charophyta

Staurastrum

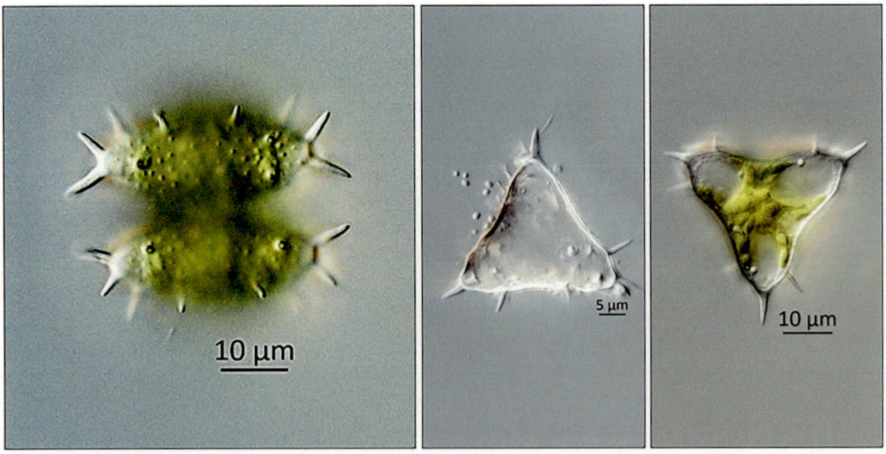

Staurastrum hystrix Metaphyton

Fig. 2.132 *Staurastrum*

Class: Zygnemataphyceae
Order: Desmidiales
Family: Desmidiaceae

Charophyta

Staurastrum

Staurastrum bieneanum *Staurastrum turgescens* *Staurastrum rotula* *Staurastrum subpygmaeum* var. *spiniferum*

Staurastrum quadrangulare

Metaphyton

Fig. 2.133 *Staurastrum*

Class: Zygnemataphyceae
Order: Desmidiales
Family: Desmidiaceae

Charophyta

Staurastrum

Staurastrum polytrichum *Staurastrum distentum*

Metaphyton

Fig. 2.134 *Staurastrum*

Class: Zygnemataphyceae
Order: Desmidiales
Family: Desmidiaceae

Charophyta

Staurastrum

Staurastrum spp.

Metaphyton

Fig. 2.135 *Staurastrum*

Class: Zygnemataphyceae
Order: Desmidiales
Family: Desmidiaceae

Charophyta

Staurodesmus

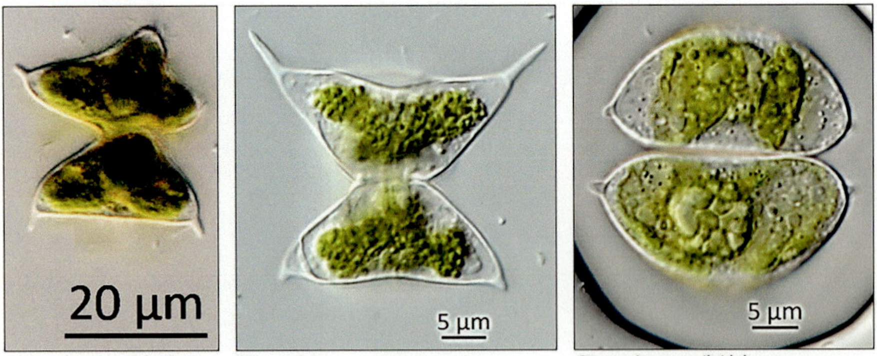

Staurodesmus dejectus *Staurodesmus omearae* *Staurodesmus dickiei*

Metaphyton

Fig. 2.136 *Staurodesmus*

Class: Zygnemataphyceae
Order: Desmidiales
Family: Desmidiaceae

Charophyta

Staurodesmus

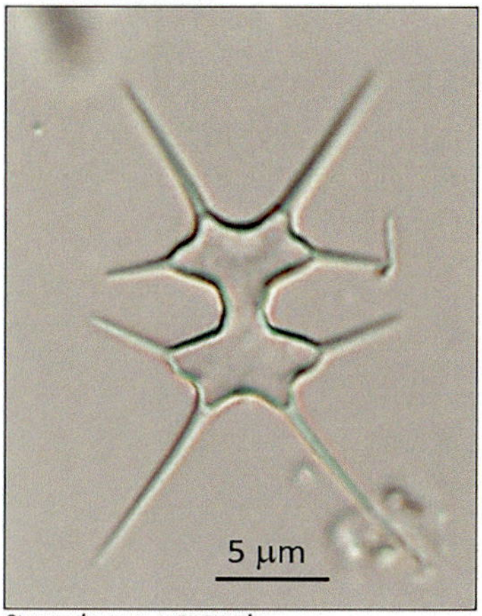

Staurodesmus octocornis

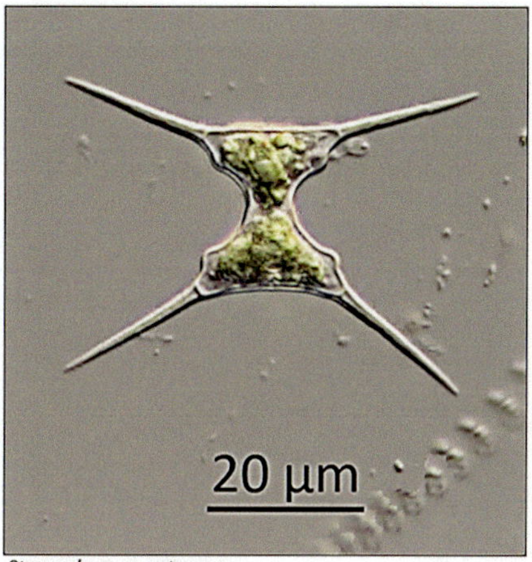

Staurodesmus extensus

Metaphyton

Fig. 2.137 *Staurodesmus*

Class: Zygnemataphyceae
Order: Desmidiales
Family: Desmidiaceae

Charophyta

Teilingia

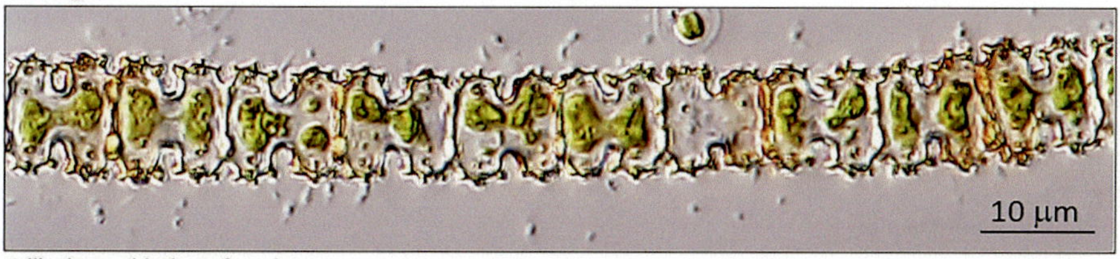

Teilingia quadrispinata f. *evoluta*

Phymatodocis

Phymatodocis nordstedtiana

Metaphyton

Fig. 2.138 *Teilingia* and *Phymatodocis*

Class: Zygnemataphyceae
Order: Desmidiales
Family: Desmidiaceae

Charophyta

Tetmemorus

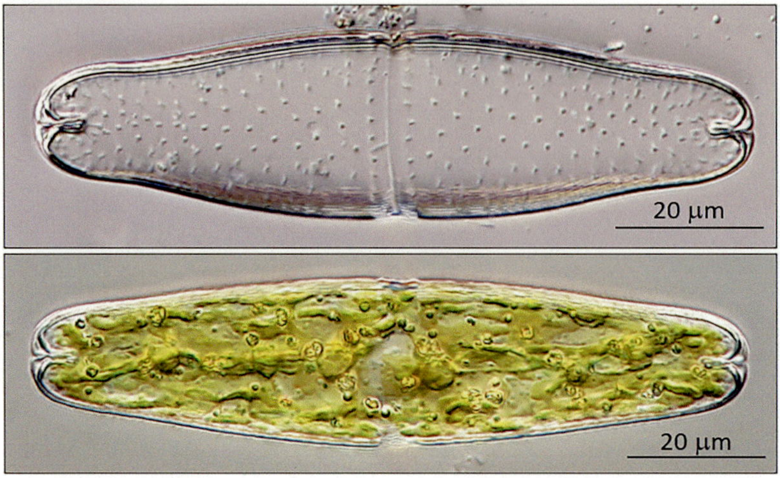

Tetmemorus granulatus

Metaphyton

Fig. 2.139 *Tetmemorus*

Class: Zygnemataphyceae
Order: Desmidiales
Family: Desmidiaceae

Charophyta

Triploceras

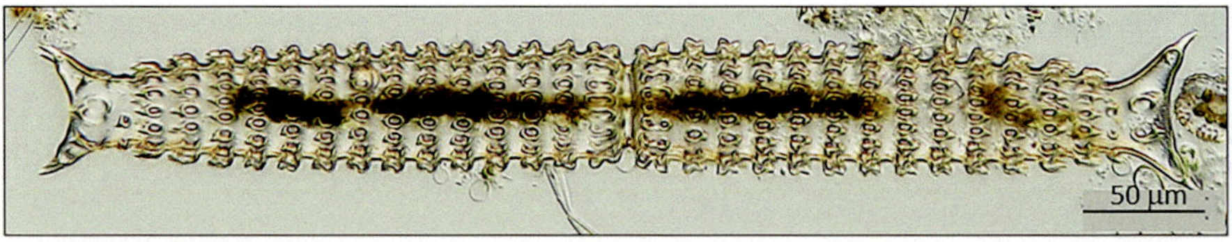

Triploceras verticillatum

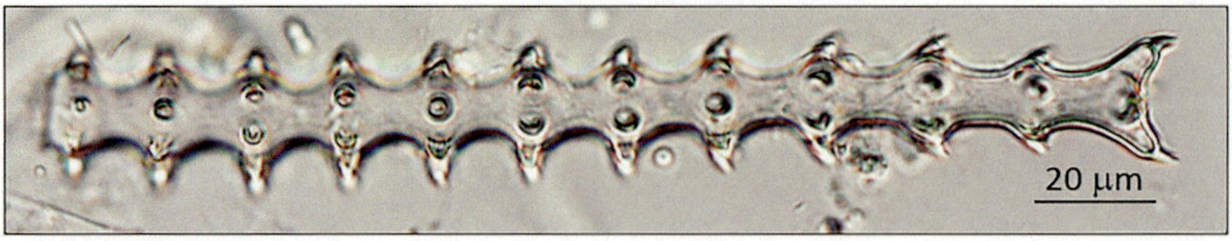

Triploceras gracile

Metaphyton

Fig. 2.140 *Triploceras*

Class: Zygnemataphyceae
Order: Desmidiales
Family: Desmidiaceae

Charophyta

Xanthidium

Xanthidium armatum

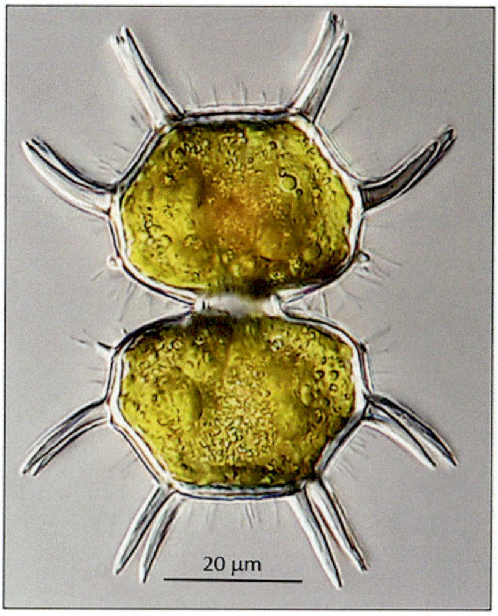

Xanthidium cristatum var. *scrobiculatum*

Metaphyton

Fig. 2.141 *Xanthidium*

Class: Zygnematophyceae
Order: Zygnematales
Family: Mesotaeniaceae

Charophyta

Netrium

Netrium digitus

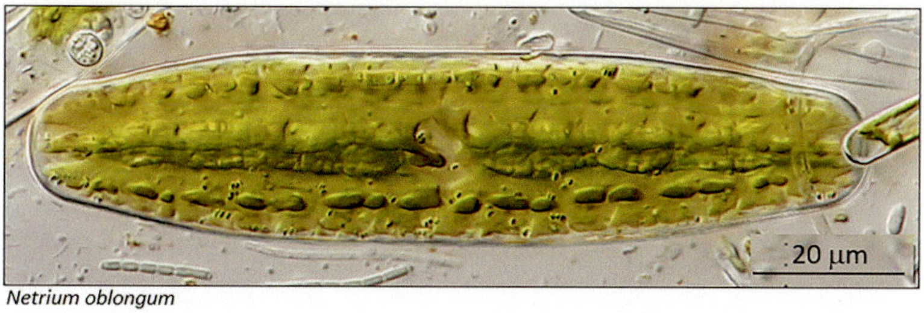

Netrium oblongum

Metaphyton

Fig. 2.142 *Netrium*

Class: Zygnematophyceae
Order: Zygnematales
Family: Mesotaeniaceae

Charophyta

Tortitaenia

Tortitaenia obscura Metaphyton

Fig. 2.143 *Tortitaenia*

Class: Zygnematophyceae
Order: Desmidiales
Family: Peniaceae

Charophyta

Penium

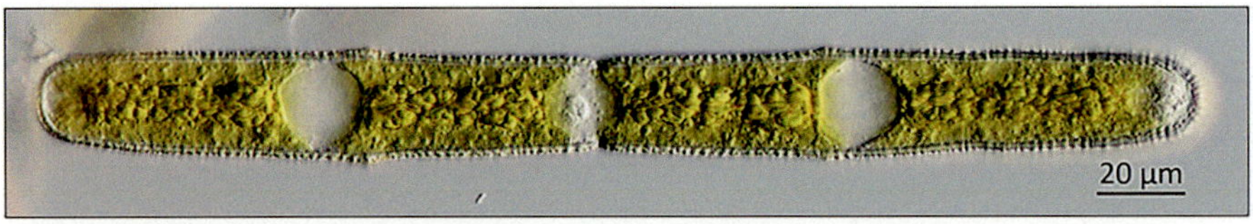

Penium margaritaceum

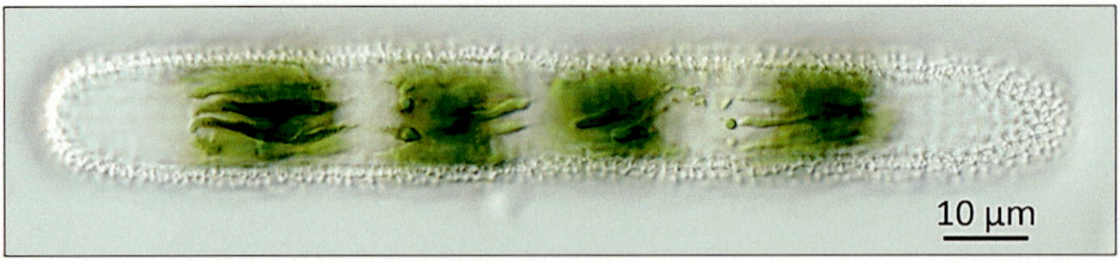

Penium sp. Metaphyton

Fig. 2.144 *Penium*

Class: Zygnemataphyceae
Order: Spirogyrales
Family: Spirogyraceae

Charophyta

Spirogyra

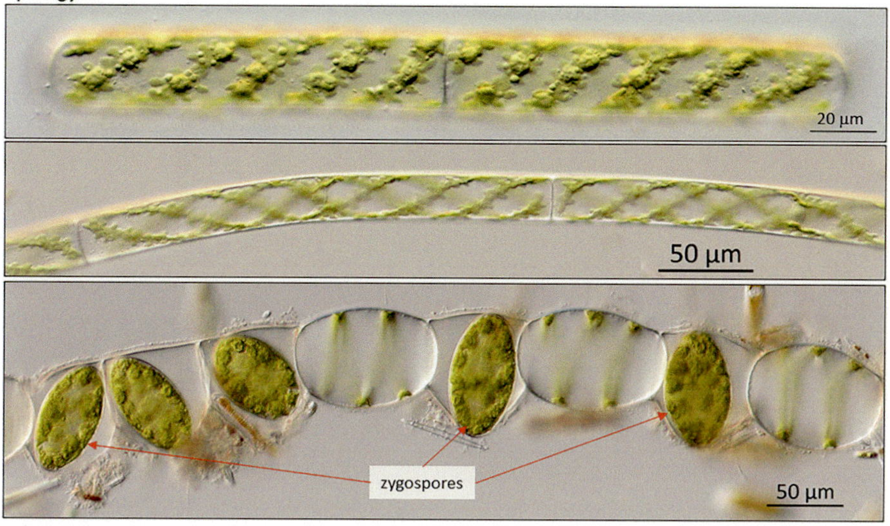

Spirogyra spp.

Fig. 2.145 *Spirogyra*

Class: Zygnemataphyceae
Order: Zynematales
Family: Zygnemataceae

Charophyta

Mougeotia

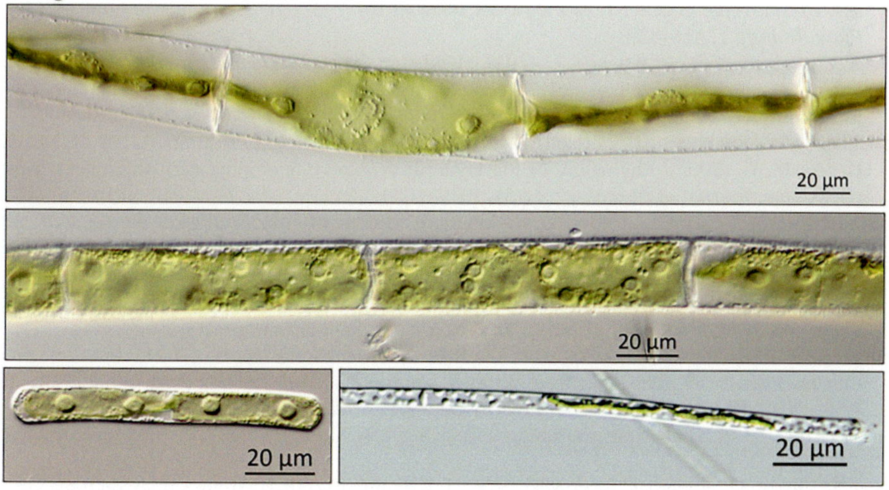

Mougeotia spp.

Metaphyton

Fig. 2.146 *Mougeotia*

Class: Zygnematophyceae
Order: Zygnematales
Family: Zygnemataceae

Charophyta

Zygnema

Zygnema spp. Metaphyton

Fig. 2.147 *Zygnema*

Suggested Reading

An, S. S., Friedl, T., & Hegewald, E. (1999). Phylogenetic relationships of *Scenedesmus* and *Scenedesmus*-like coccoid green algae as inferred from IT-2 rDNA sequence comparisons. *Plant Biology, 1*, 418–428.

Bock, C., Pröschold, T., & Krienitz, L. (2011). Updating the genus *Dictyosphaerium* and description of *Mucidosphaerium* gen. nov. (Trebouxiophyceae) based on morphological and molecular data. *Journal of Phycology, 47*(3), 638–652.

Buchheim, M., Buchheim, J., Carlson, T., Braband, A., HepperLe, D., Krienitz, L., Wolf, M., & Hegewald, E. (2005). Phylogeny of the Hydrodictyaceae (Chlorophyceae): Inferences from rDNA data. *Journal of Phycology, 41*(5), 1039–1054.

Cavalier-Smith, T. (1998). A revised six-kingdom system of life. *Biological Reviews of the Cambridge Philosophical Society, 73*, 203–266.

Croasdale, H. T., Bicudo, C. E. M., & Prescott, G. W. (1983). *A synopsis of North American desmids. Part II. Desmidiaceae: Placodermae section 5. The filamentous genera* (pp. vii, 117, plates 448–470). University of Nebraska Press.

Fuciková, K., Williamson, V., Choquette, C., Bustos, D., & Pietrasiak, N. (2024). First insights into the diversity of green algae from White Sands National Park and Organ Mountains–Desert Peaks National Monument. *Western North American Naturalist, 84*(1), 107–124.

Hindák, F. (1988). Studies on the chlorococcal algae (Chlorophyceae) IV. *Biologické Práce, 34*(1/2), 1–264.

Jena, M., Bock, C., Behera, C., Adhikary, S. P., & Krienitz, L. (2014). Strain survey on three continents confirms the polyphyly of the genus *Pediastrum* (Hydrodictyaceae, Chlorophyceae). *Fottea, 14*(1), 63–76.

John, D. M., Wynne, M. J., & Tsarenko, P. M. (2014). Reinstatement of the genus *Willea* Schmidle 1900 for *Crucigeniella* Lemmermann 1900 *nom. illeg.* (Chlorellales, Trebouxiophyceae, Chlorophyta). *Phytotaxa, 167*(2), 212–214.

Komárek, J. (1979). Änderungen in der Taxonomie der Chlorokokkalalgen. *Archiv für Hydrobiologie Suppl. 56 (Algological Studies), 24*, 239–263.

Komárková-Legnerová, J. (1969). The systematics and ontogenesis of the genera *Ankistrodesmus* Corda and *Monoraphidium* gen. nov. In B. Fott (Ed.), *Studies in phycology* (pp. 75–144).

Kützing, F. T. (1843). *Phycologia generalis* oder Anatomie, Physiologie und Systemkunde der Tange. Mit 80 farbig gedruckten Tafeln, gezeichnet und gravirt vom Verfasser. (pp. [part 1]: [i]–xxxii, [1]-142, [part 2:] 143–458, 1, err.], pls 1–80). Leipzig: F.A. Brockhaus.

Liu, B. W., Hu, Y. X., Hu, Z. Y., Liu, G. X., & Zu, H. (2020). Taxonomic scheme of the order Chaetophorales (Chlorophyceae, Chlorophyta) based on chloroplast genomes. *BMC Genomics, 21*(442), 1–12.

Nägeli, C. (1849). Gattungen einzelliger Algen, physiologisch und systematisch bearbeitet. *Neue Denkschriften der Allg. Schweizerischen Gesellschaft für die Gesammten. Naturwissenschaften, 10*(7), i–viii, 1–139, pls I–VIII.

Prescott, G. W. (1951). Algae of the Western Great Lakes area exclusive of desmids and diatoms. *Bulletin of the Cranbrook Institute of* Science No. 31: [i]–xiii, 1–946.

Prescott, G. W., Croasdale, H. T., & Vinyard, W. C. (1975). *A synopsis of North American desmids. Part II. Desmidiaceae: Placodermae. Sect. 1* (pp. [i–vii], 1–275, incl. 57 pls). University of Nebraska Press.

Prescott, G. W., Croasdale, H. T., & Vinyard, W. C. (1977). *A synopsis of North American desmids part II. Desmidiaceae: Placodermae section 2* (pp. [i-vii], 1–411, incl. Pls LVIII-CXLVII). University of Nebraska Press.

Prescott, G. W., Croasdale, H. T., Vinyard, W. C., & Bicudo, C. E. (1981). *A synopsis of north American desmids. Part II. Desmidiaceae: Placodermae section 3* (pp. vii, 720, plates 148–293). University of Nebraska Press.

Rosen, B. H., Stahlhut, K. N. & Hall, J. D. (2019). Catalog of microscopic organisms of the Everglades, Part 2 – The desmids of the Arthur R. Marshall Loxahatchee National Wildlife Refuge. *U.S. Geological Survey Scientific Investigations Report* 2019–5074.

Smith, G. M. (1913). *Tetradesmus*, a new four-celled coenobic alga. *Bulletin of the Torrey Botanical Club, 40*, 75–87.

Smith, G. M. (1950). *The fresh-water algae of the United States* (pp. [i]-vii, [1]-719, 559 figs) (2nd ed.). McGraw-Hill Book Company.

Wehr, J., Sheath, R., & Kociolek, J. P. (2015). *Freshwater algae of North America: Ecology and classification* (pp. [1]-xvi, 1-1050) (2nd ed.). Elsevier, Academic Press.

West, W. & West, G.S. (1904). *A monograph of the British Desmidiaceae. I* (pp. [i]–xxxvi, [1]–224, pl. 1–32).

Whitford, L., & Schumacher, G. (1984). *A manual of fresh-water algae in North Carolina* (p. 337). Sparks Press.

Diatoms

3

Diatoms, a group of microscopic eukaryotic algae, are vital components of aquatic ecosystems worldwide. These single-celled organisms belong to the class Bacillariophyceae within the kingdom Chromista. Diatoms' most distinctive feature is their cell wall, which is composed of silica. This silica-based cell wall, known as a frustule, is composed of two halves, known as valves. The valves have an intricate pattern of pores, ridges, and other features that allow nutrients and gases to be exchanged with their surroundings. Reproduction is achieved by simply forming two new cells within the parent frustule, leading to a reduction in cell size with each division. Sexual reproduction triggers the formation of an enlarged cell, the auxospore that restores the full-size cells.

The identification of diatoms is mostly based on the arrangement of these features, some of which are only visible with an electron microscope. Diatoms are divided into several groups. The centric diatoms have valves that are mostly radially symmetrical and are not motile. To keep suspended in the water column, many centric diatoms form long chains of cells (see *Aulacoseira,* Fig. 3.2) or produce chitin fibrils (see *Stephanodiscus*, Fig. 3.3). The araphid diatoms, some of which are bilaterally symmetrical, are also nonmotile but commonly secrete a mucilaginous stalk for attaching rocks, sand, or other organisms. The third group has a slit (raphe) down the middle of one valve that provides some motility on a substrate. The fourth group has a raphe on both valves, or up on a raised keel.

Diatoms contain a variety of pigments, including chlorophyll *a* and *c*, as well as carotenoids like fucoxanthin. These pigments enable diatoms to capture sunlight for photosynthesis from various depths in the water column. Fucoxanthin gives diatoms their characteristic golden-brown color. Diatoms are incredibly abundant and diverse, inhabiting a wide range of aquatic environments, from freshwater lakes and rivers to marine systems. Diatoms contribute significantly to the Earth's primary productivity, generating a substantial portion of the oxygen we breathe.

Diatoms play a crucial role in aquatic food webs. As primary producers, they form the base of the food web, providing sustenance for a variety of organisms including benthic aquatic insects that graze on the attached diatoms and

B. H. Rosen, *Color Atlas of Freshwater Algae*,
https://doi.org/10.1007/978-3-031-93122-2_3

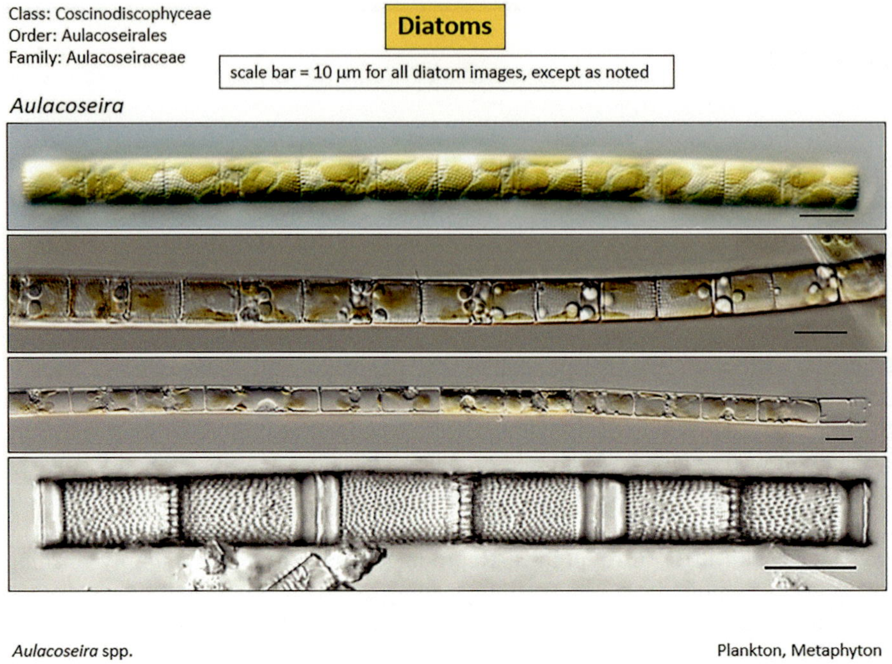

Fig. 3.1 *Aulacoseira*

filter-feeding zooplankton. They can also be used as an indicator species as some have a narrow preference range for environmental variables, such as pH (Figs. 3.1 to 3.74).

Class: Coscinodiscophyceae
Order: Stephanodiscales
Family: Stephanodiscaeae

Diatoms

Stephanodiscus

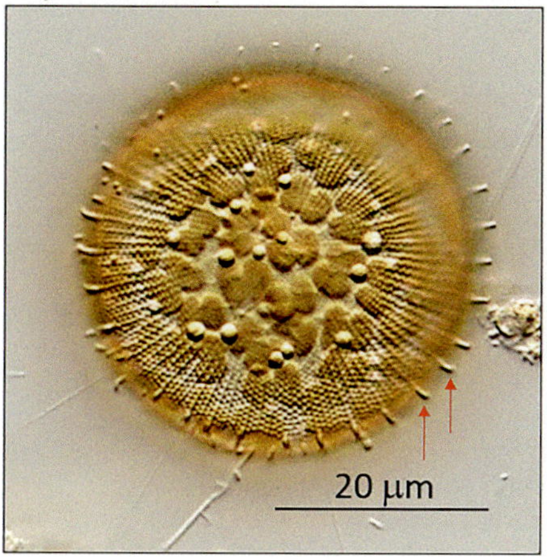

spines (at arrows)

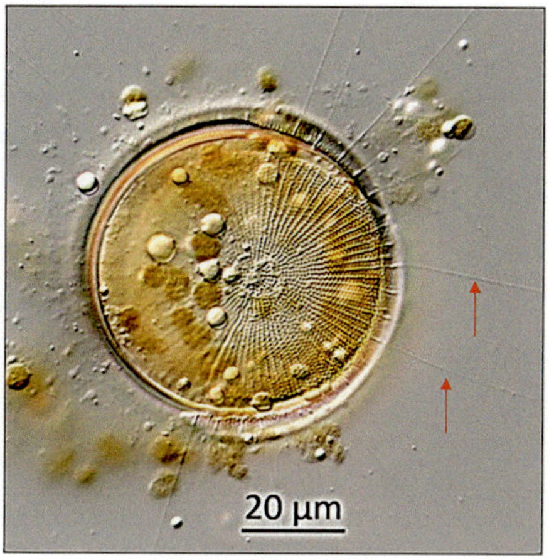

chitin fibrils (at arrows)

Stephanodiscus sp.

Plankton

Fig. 3.2 *Stephanodiscus*

Class: Coscinodiscophyceae
Order: Thalassiosirales
Family: Melosiraceae

Diatoms

scale bar = 10 µm for all diatom images, except as noted

Melosira

Melosira varians C.Agardh 1827: 628. https://diatoms.org/genera/melosira

fresh water

Melosira nummuloides

brackish waters

Plankton, Metaphyton

Fig. 3.3 *Melosira*

Class: Coscinodiscophyceae
Order: Cosinodiscales
Family: Hemidiscaeae

Diatoms

Actinocyclus

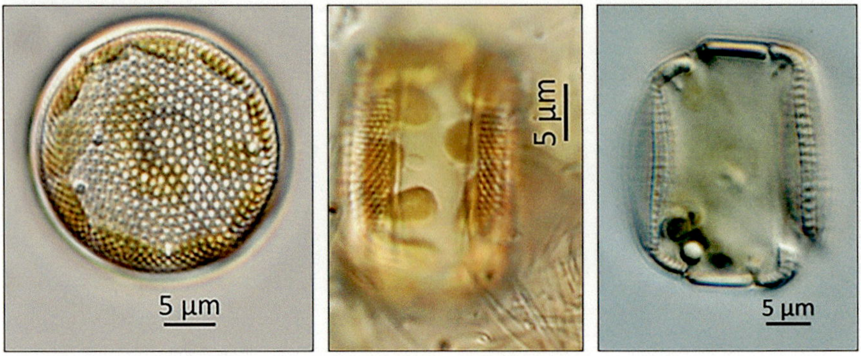

Actinocyclus normanii

Plankton

Fig. 3.4 *Actinocyclus*

Class: Coscinodiscophyceae
Order: Thalassiosirales
Family: Stephanodiscaeae

Diatoms

Cyclotella

Cyclotella spp.

Plankton

Fig. 3.5 *Cyclotella*

Class: Coscinodiscophyceae
Order: Thalassiosirales
Family: Stephanodiscaeae

Diatoms

Stephanocyclus

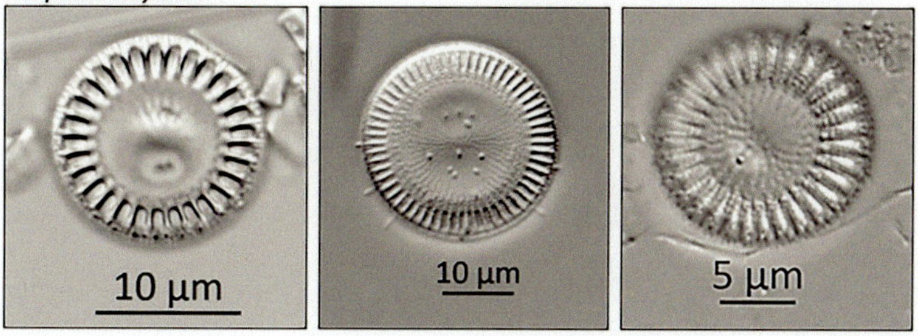

Stephanocyclus spp.

Plankton

Fig. 3.6 *Stephanocyclus*

Class: Coscinodiscophyceae
Order: Thalassiosirales
Family: Stephanodiscaeae

Diatoms

Discostella

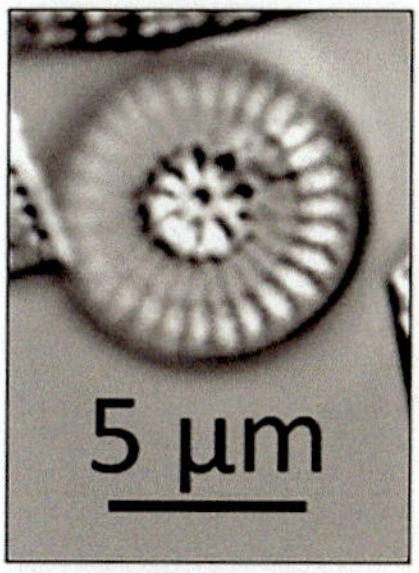

Discostella sp.

Plankton

Fig. 3.7 *Discostella*

Class: Mediophyceae
Order: Eupodiscales
Family: Eupodiscaceae

Diatoms

Pleurosira

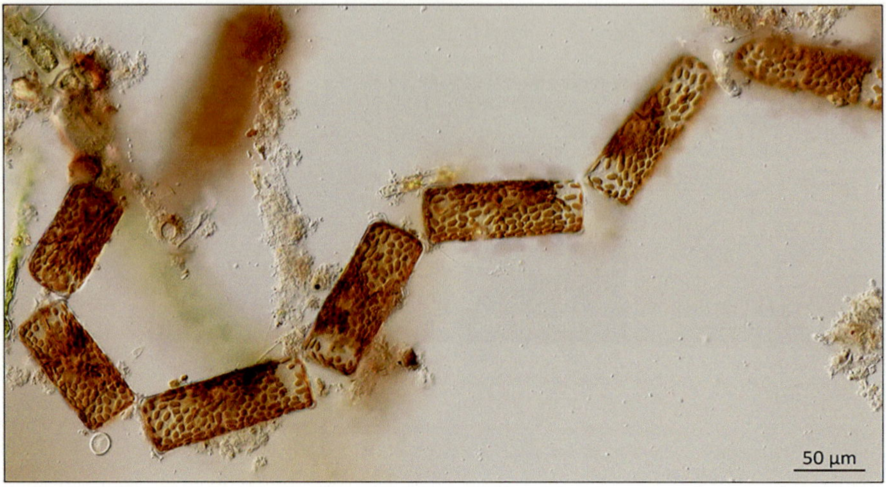

Pleurosira laevis — Epiphyton, Benthic

Fig. 3.8 *Pleurosira*

Class: Fragilariophyceae
Order: Fragilariales
Family: Fragilariaceae

Diatoms

scale bar = 10 µm for all diatom images, except as noted

Asterionella

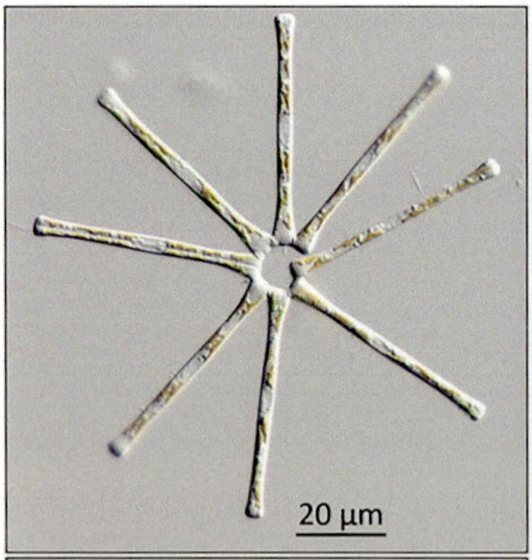

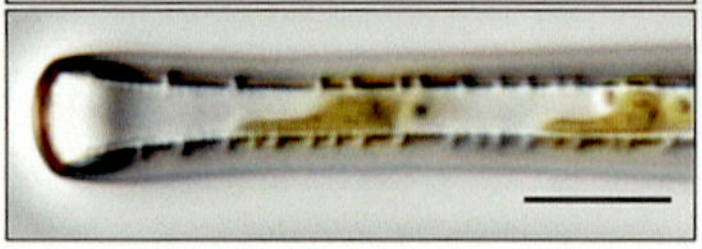

chytrid (fungi, at arrows)

Asterionella formosa — Plankton

Fig. 3.9 *Asterionella*

Class: Fragilariophyceae
Order: Fragilariales
Family: Fragilariaceae

Diatoms

scale bar = 10 mm for all diatom images, except as noted

Diatoma

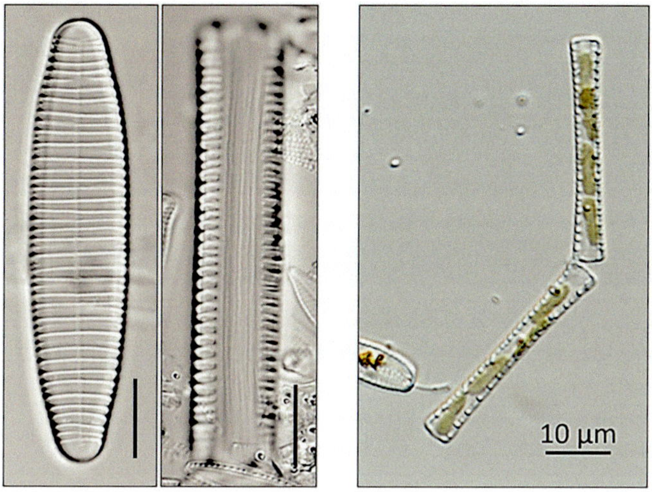

Diatoma vulgaris

Epiphyton, epilithic

Fig. 3.10 *Diatoma*

Class: Fragilariophyceae
Order: Fragilariales
Family: Fragilariaceae

Diatoms

scale bar = 10 mm for all diatom images, except as noted

Meridion

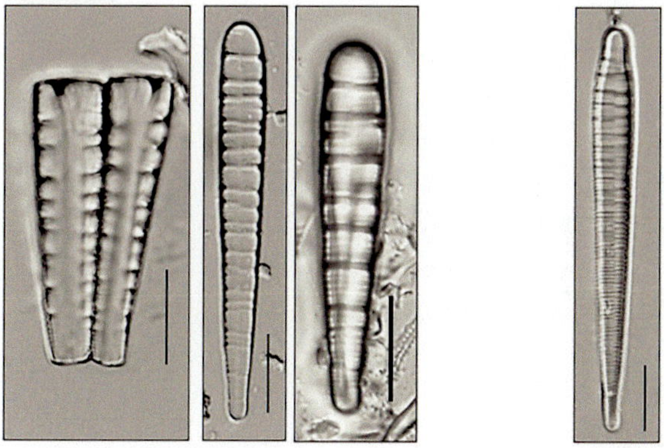

Meridion circulare

Meridion circulare
var. *contrictum*

Epiphyton, epilithic

Fig. 3.11 *Meridion*

Class: Fragilariophyceae
Order: Fragilariales
Family: Fragilariaceae

Diatoms

scale bar = 10 μm for all diatom images, except as noted

Ulnaria

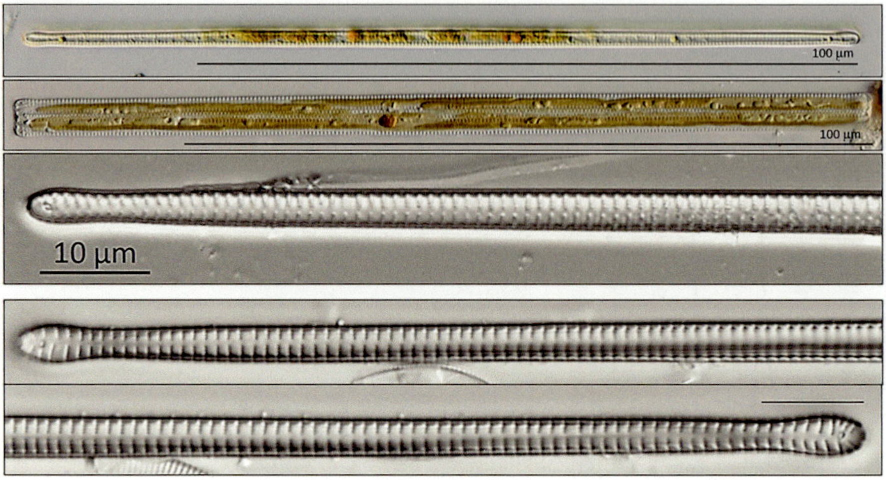

Ulnaria delicatissima Plankton, epiphyton

Fig. 3.12 *Ulnaria*

Class: Fragilariophyceae
Order: Fragilariales
Family: Fragilariaceae

Diatoms

Fragilaria

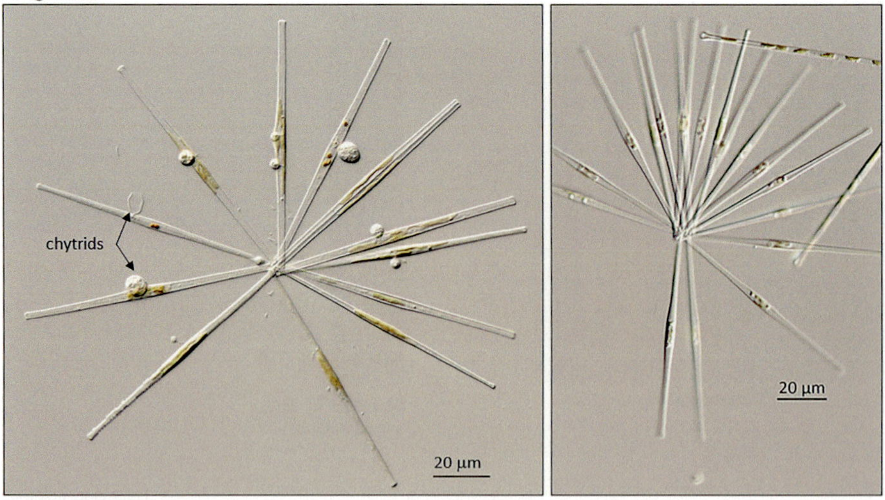

Fragilaria lemanenesis Plankton

Fig. 3.13 *Fragilaria*

Class: Fragilariophyceae
Order: Fragilariales
Family: Fragilariaceae

Diatoms

scale bar = 10 μm for all diatom images, except as noted

Fragilaria

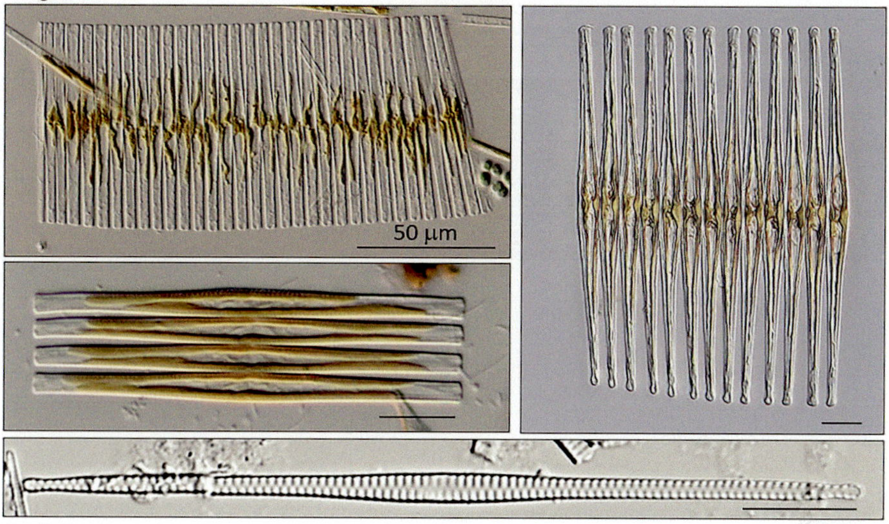

Fragilaria crotonensis

Plankton

Fig. 3.14 *Fragilaria*

Class: Fragilariophyceae
Order: Fragilariales
Family: Fragilariaceae

Diatoms

Fragilaria

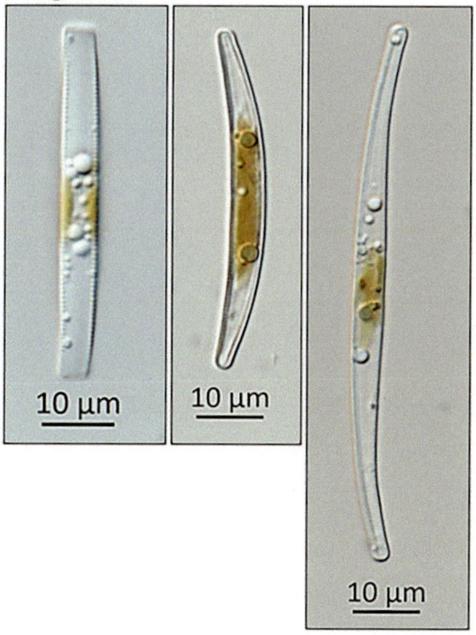

Fragilaria cyclopum

Plankton

Fig. 3.15 *Fragilaria*

Class: Fragilariophyceae
Order: Fragilariales
Family: Fragilariaceae

Diatoms

scale bar = 10 μm for all diatom images, except as noted

Fragilaria

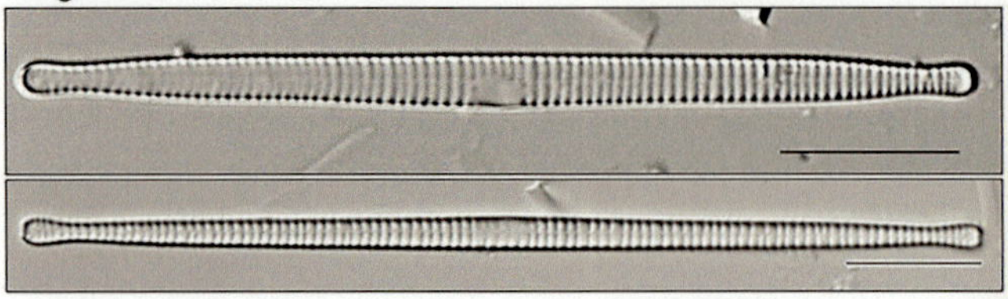

Fragilaria synegrotesca

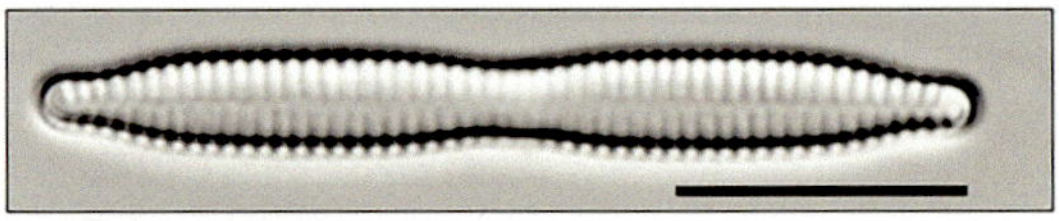

Fragilaria mesolepta

Periphyton

Fig. 3.16 *Fragilaria*

Class: Fragilariophyceae
Order: Fragilariales
Family: Fragilariaceae

Diatoms

Hannaea

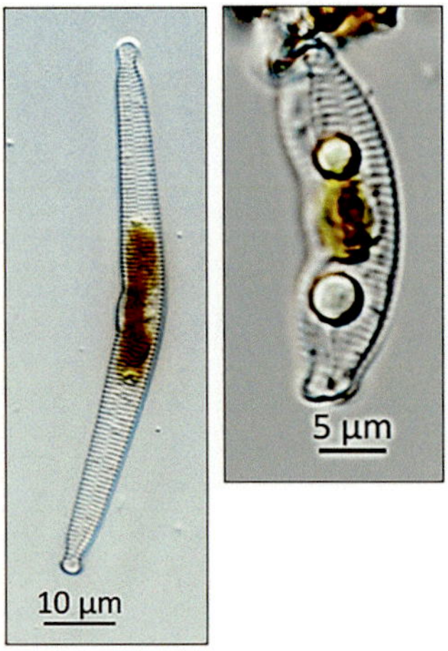

Hannaea arcus

Epilithic, periphyton

Fig. 3.17 *Hannaea*

Class: Fragilariophyceae
Order: Fragilariales
Family: Fragilariaceae

Diatoms

Staurosira

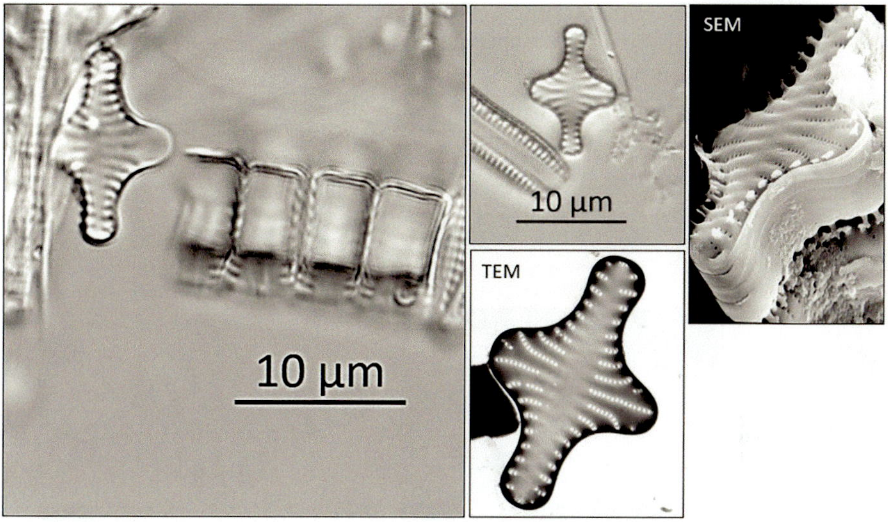

Staurosira construens

Epilithic, periphyton

Fig. 3.18 *Staurosira*

Class: Fragilariophyceae
Order: Fragilariales
Family: Fragilariaceae

Diatoms

scale bar = 10 µm for all diatom images, except as noted

Staurosira

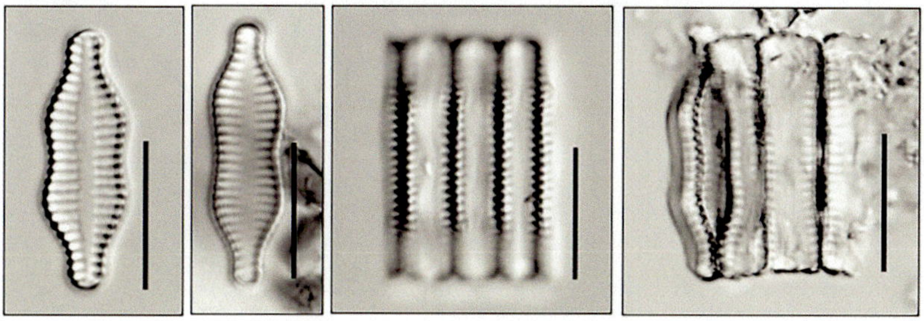

Staurosira binodis

Epilithic, periphyton

Fig. 3.19 *Staurosira*

Class: Fragilariophyceae
Order: Fragilariales
Family: Fragilariaceae

Diatoms

scale bar = 10 μm for all diatom images, except as noted

Staurosirella

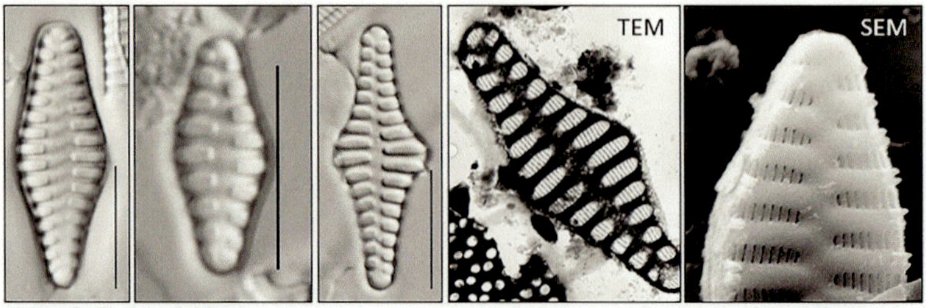

Staurosirella rhomboides

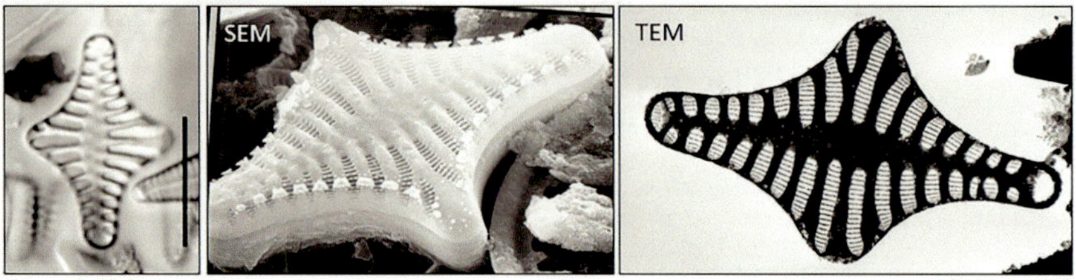

Staurosirella leptostauron

Epilithic, periphyton

Fig. 3.20 *Staurosirella*

Class: Fragilariophyceae
Order: Fragilariales
Family: Fragilariaceae

Diatoms

scale bar = 10 μm for all diatom images, except as noted

Staurosirella

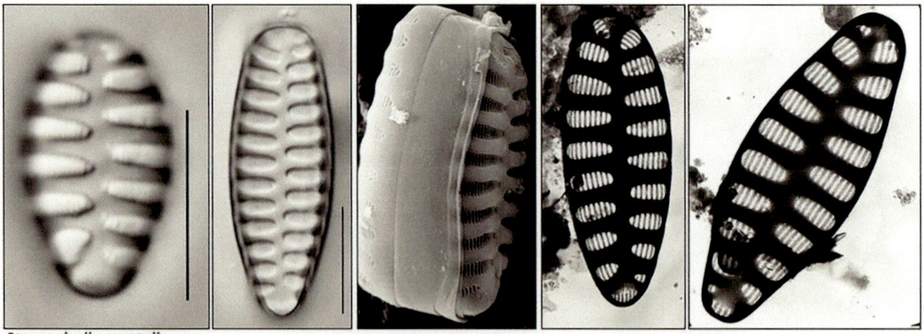

Staurosirella martyii

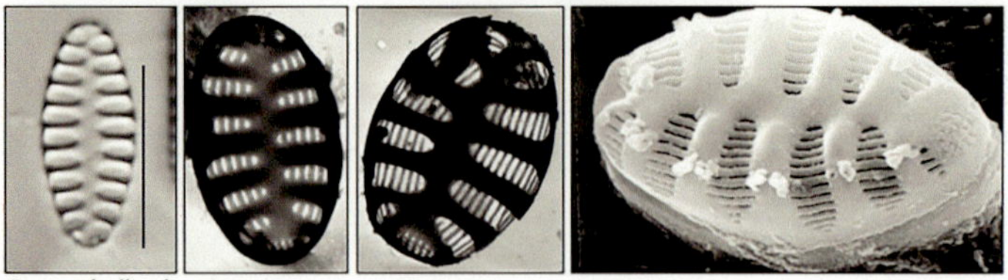

Staurosirella pinnata

Epilithic, periphyton

Fig. 3.21 *Staurosirella*

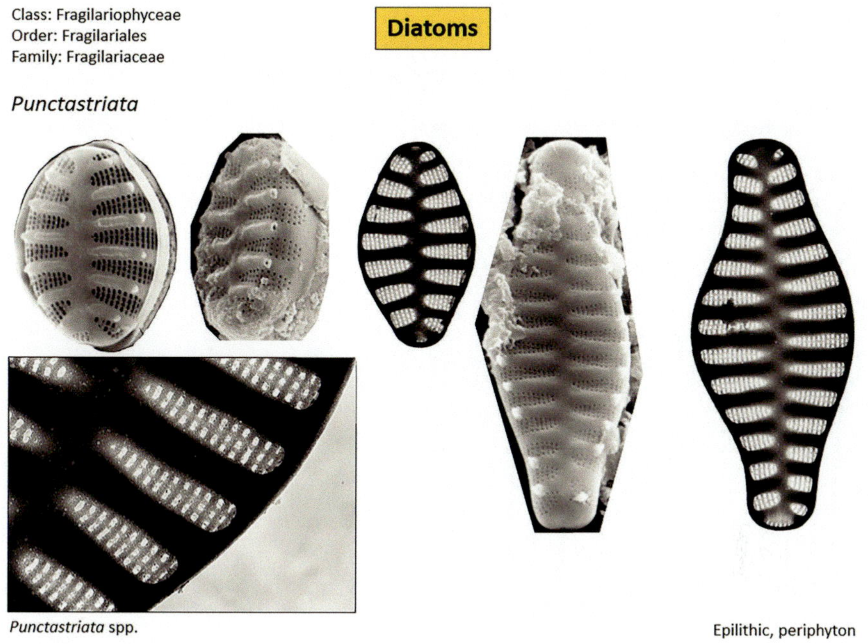

Fig. 3.22 *Punctastriata*

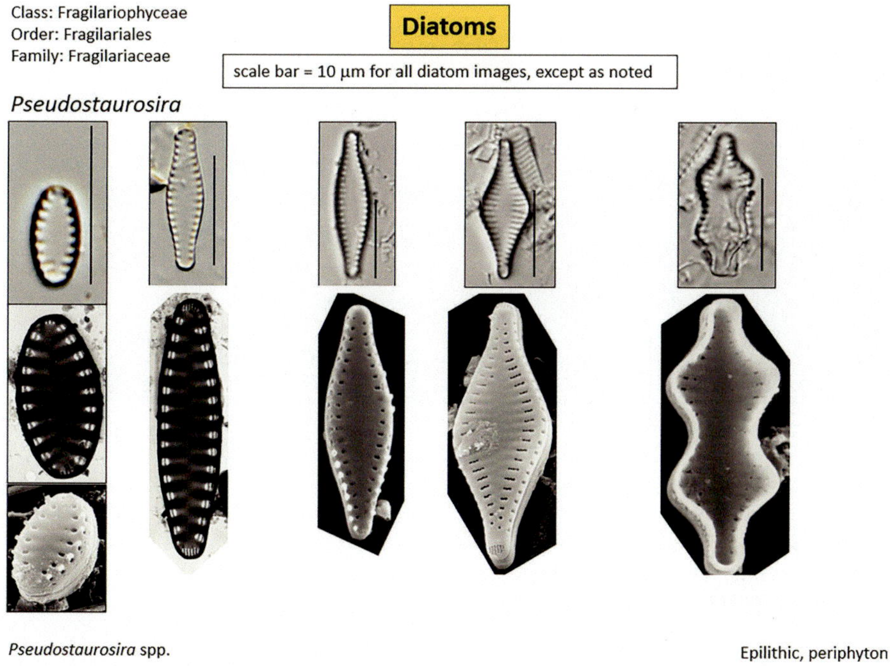

Fig. 3.23 *Pseudostaurosira*

Class: Fragilariophyceae
Order: Tabellariales
Family: Tabellareaceae

Diatoms

scale bar = 10 µm for all diatom images, except as noted

Tabellaria

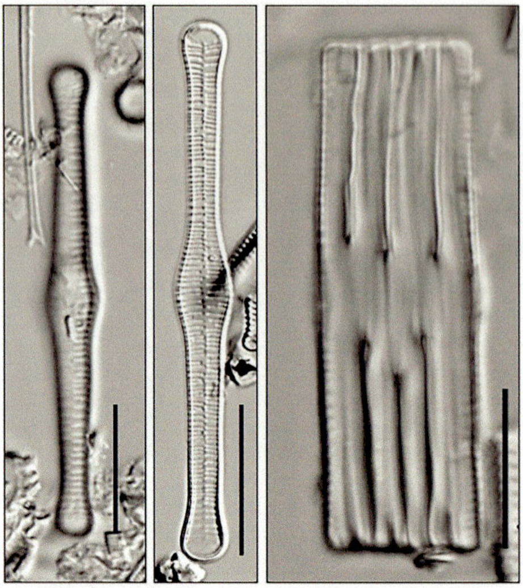

Tabellaria sp. Periphyton, plankton

Fig. 3.24 *Tabellaria*

Class: Bacillariophyceae
Order: Achnanthales
Family: Achnanthaceae

Diatoms

Achnanthes

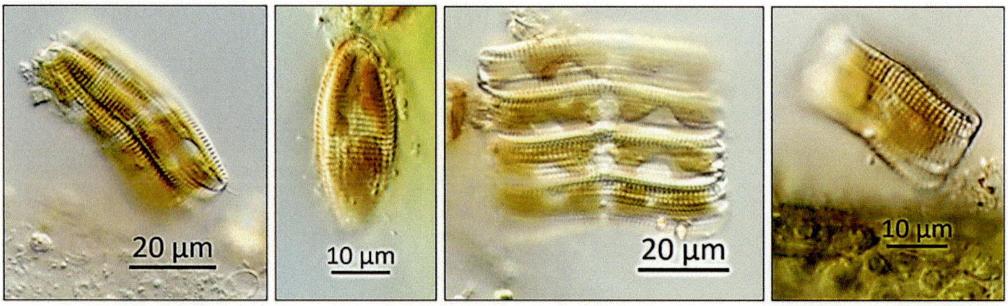

Achnanthes sp. Marine-influenced canal Periphyton, epiphyton

Fig. 3.25 *Achnanthes*

Class: Bacillariophyceae
Order: Achnanthales
Family: Achnanthaceae

Diatoms

scale bar = 10 μm for all diatom images, except as noted

Achnanthidium

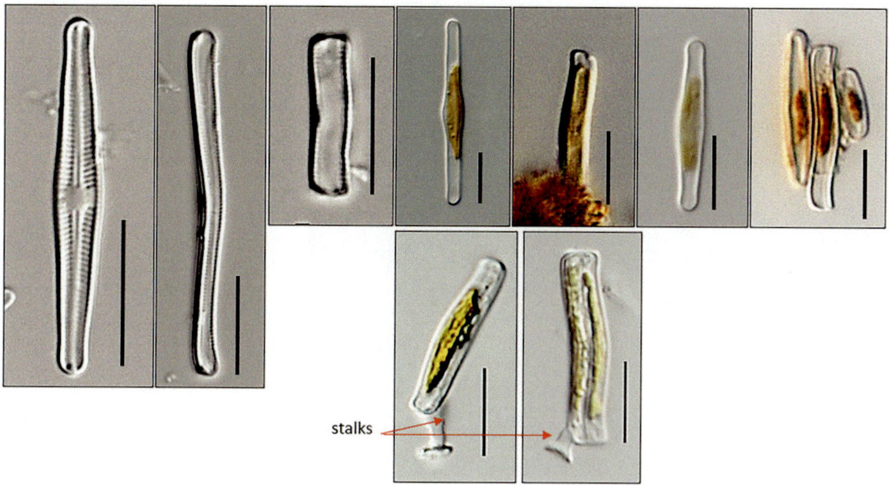

Achnanthidium spp.

Periphyton, epiphyton

Fig. 3.26 *Achnanthidium*

Class: Bacillariophyceae
Order: Achnanthales
Family: Achnanthidiaceae

Diatoms

scale bar = 10 μm for all diatom images, except as noted

Planothidium

Planothidium spp.

Periphyton, epiphyton

Fig. 3.27 *Planothidium*

Class: Bacillariophyceae
Order: Achnanthales
Family: Achnanthidiaceae

Diatoms

scale bar = 10 μm for all diatom images, except as noted

Gogorevia

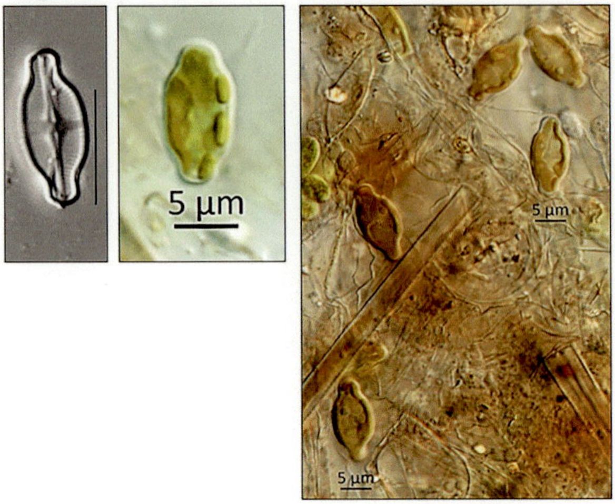

Gogorevia exilis

Periphyton, epiphyton

Fig. 3.28 *Gogorevia*

Class: Bacillariophyceae
Order: Achnanthales
Family: Cocconeidaceae

Diatoms

scale bar = 10 μm for all diatom images, except as noted

Cocconeis

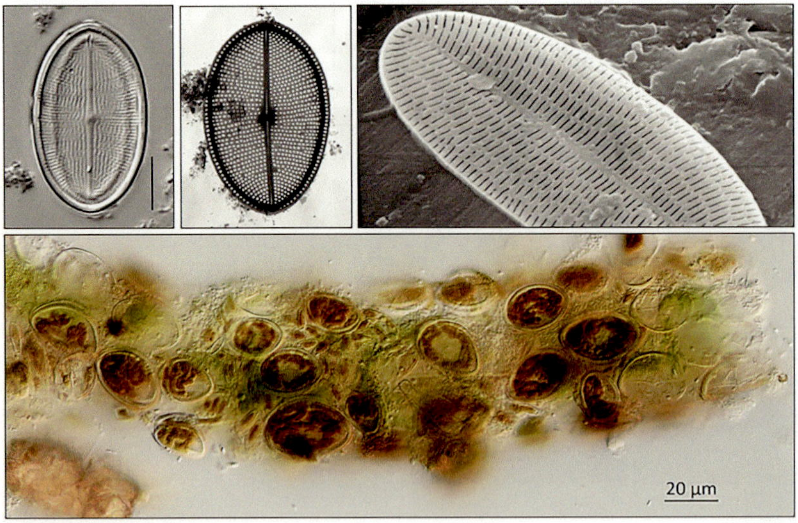

Cocconeis spp.

Periphyton, epiphyton

Fig. 3.29 *Cocconeis*

Class: Bacillariophyceae
Order: Eunotiales
Family: Eunotiaceae

Diatoms

Eunotia

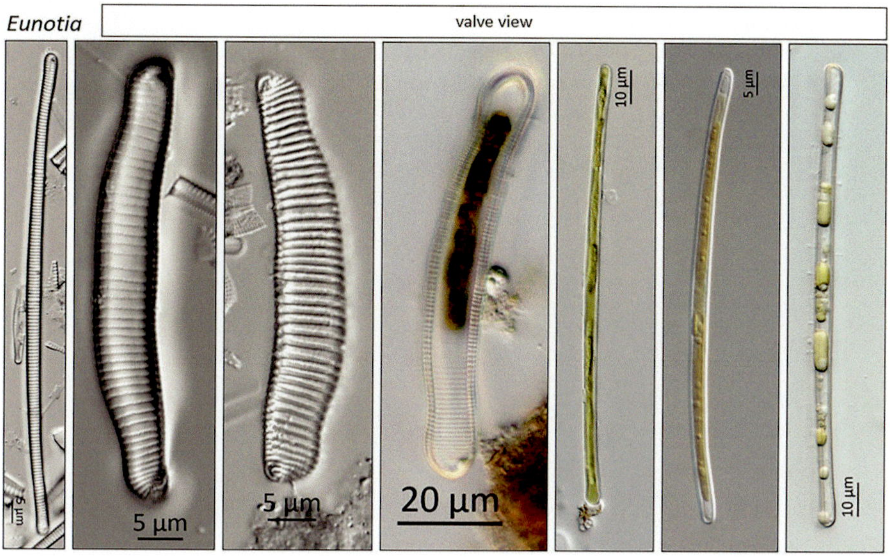

Eunotia spp.

Periphyton, epiphyton

Fig. 3.30 *Eunotia*

Class: Bacillariophyceae
Order: Eunotiales
Family: Eunotiaceae

Diatoms

Eunotia

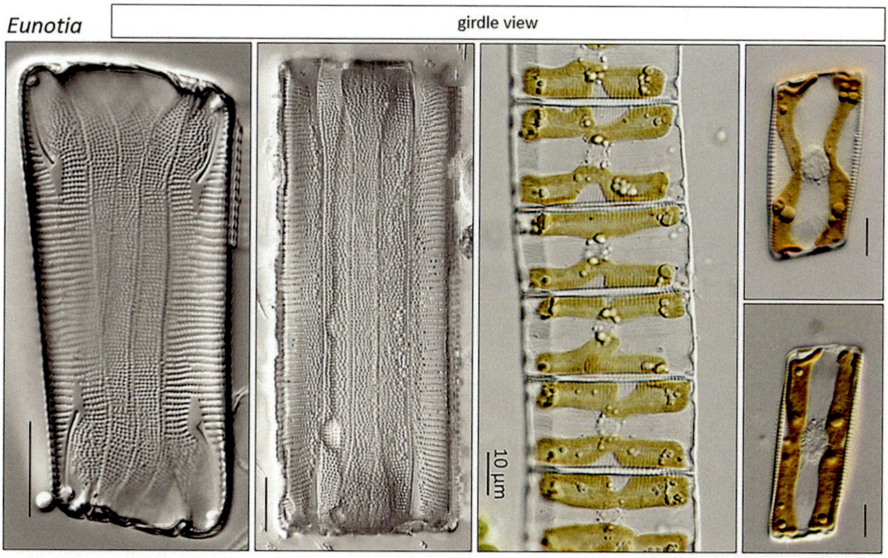

Eunotia spp.

Periphyton, epiphyton

Fig. 3.31 *Eunotia*

Class: Bacillariophyceae
Order: Eunotiales
Family: Eunotiaceae

Diatoms

scale bar = 10 µm for all diatom images, except as noted

Desmogonium

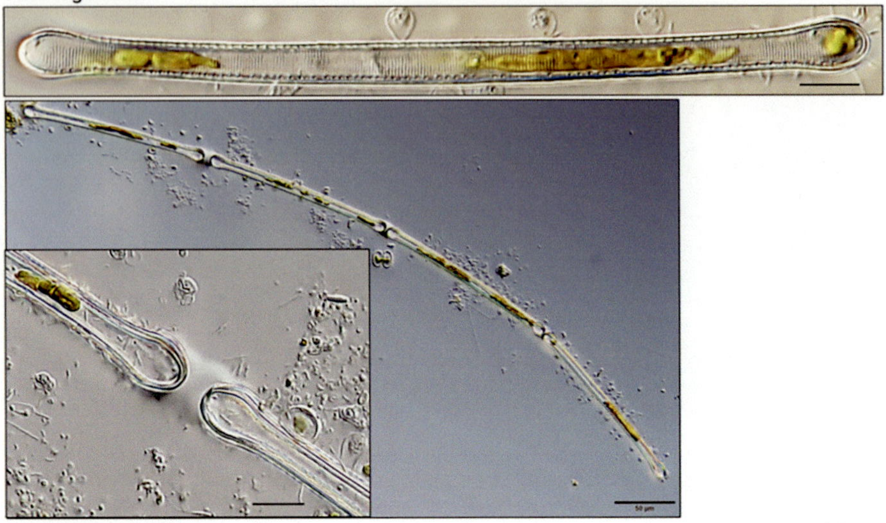

Desmogonium femoriforme Epiphyton

Fig. 3.32 *Desmogonium*

Class: Bacillariophyceae
Order: Naviculales
Family: Naviculaceae

Diatoms

scale bar = 10 µm for all diatom images, except as noted

Navicula

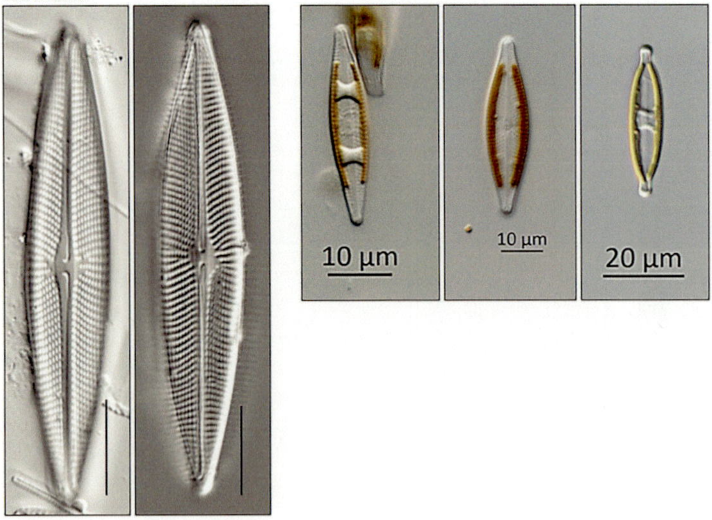

Navicula spp. Periphyton

Fig. 3.33 *Navicula*

Class: Bacillariophyceae
Order: Naviculales
Family: Stauroneidaceae

Diatoms

scale bar = 10 µm for all diatom images, except as noted

Stauroneis

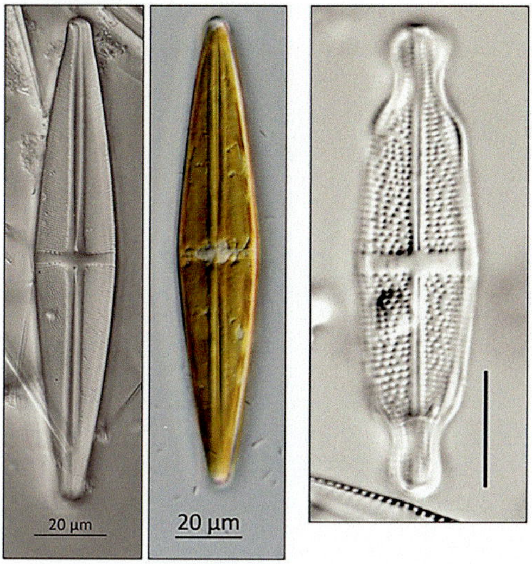

Stauroneis spp.

Periphyton

Fig. 3.34 *Stauroneis*

Class: Bacillariophyceae
Order: Naviculales
Family: Stauroneidaceae

Diatoms

Craticula

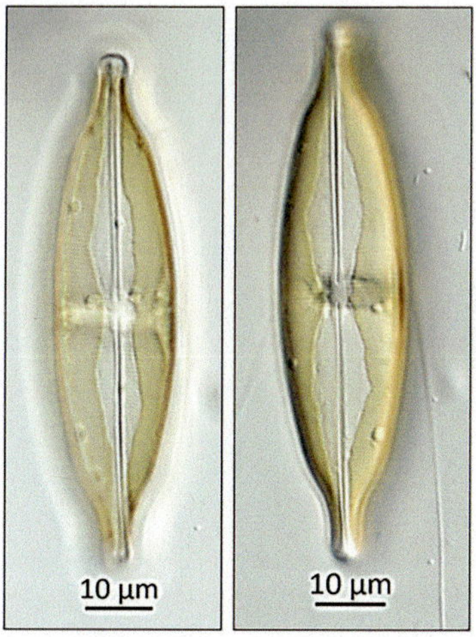

Craticula riparia

Periphyton

Fig. 3.35 *Craticula*

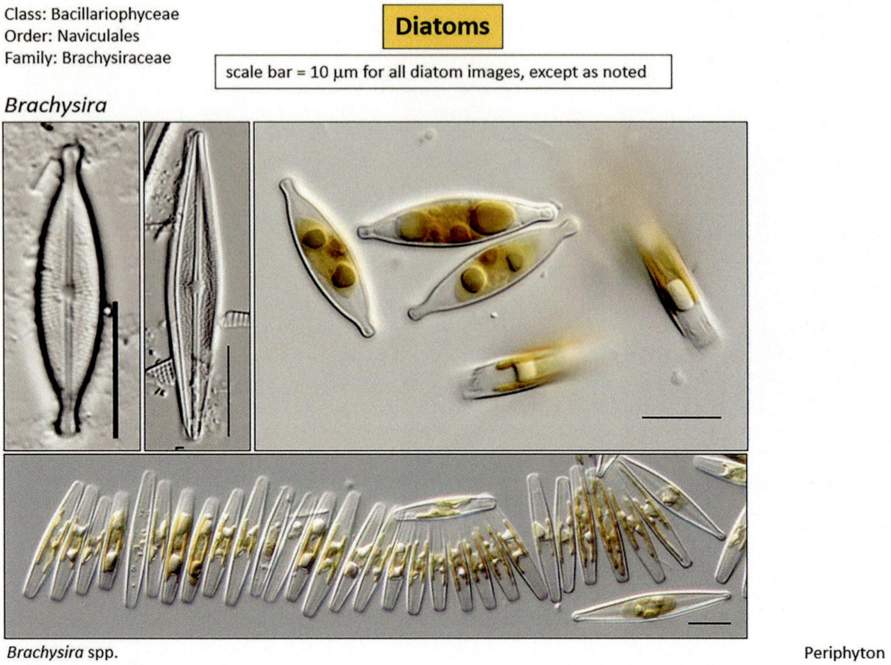

Fig. 3.36 *Brachysira*

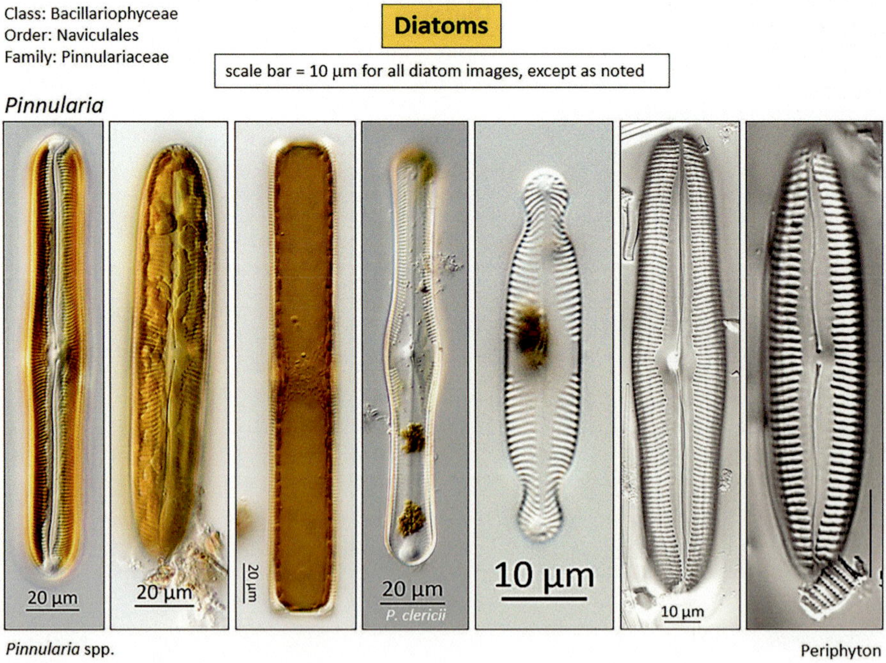

Fig. 3.37 *Pinnularia*

Class: Bacillariophyceae
Order: Naviculales
Family: Pinnulariaceae

Diatoms

Pinnularia

Pinnularia sp. *Pinnularia borealis*

Periphyton

Fig. 3.38 *Pinnularia*

Class: Bacillariophyceae
Order: Naviculales
Family: Sellaphoraceae

Diatoms

scale bar = 10 µm for all diatom images, except as noted

Sellaphora

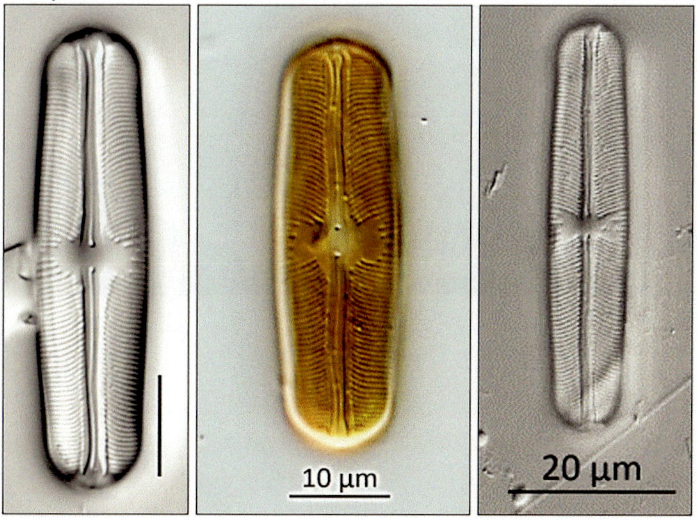

Sellaphora spp.

Periphyton

Fig. 3.39 *Sellaphora*

Class: Bacillariophyceae
Order: Naviculales
Family: Sellaphoraceae

Diatoms

Fallacia (cf)

Fallacia cf. *hodgeana* Periphyton

Fig. 3.40 *Fallacia* (cf)

Class: Bacillariophyceae
Order: Mastogloiales
Family: Mastogloiaceae

Diatoms

scale bar = 10 µm for all diatom images, except as noted

Aneumastus

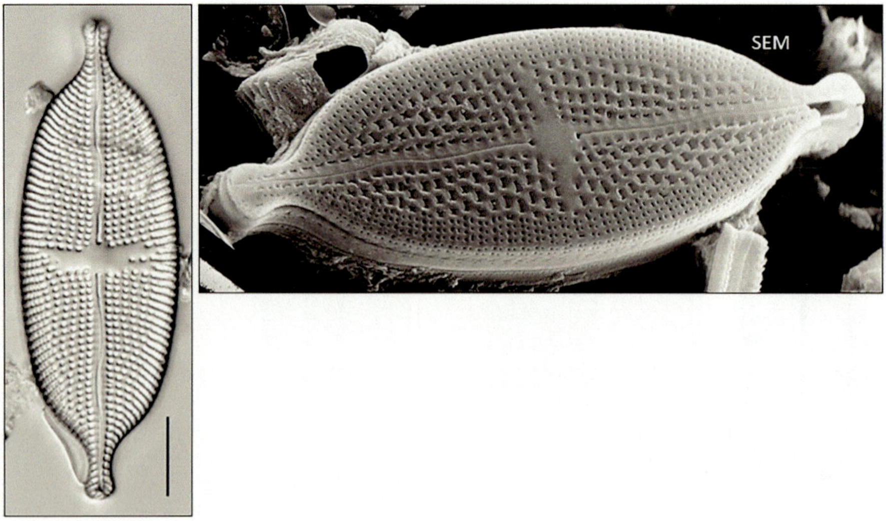

Aneumastus tusculus Periphyton

Fig. 3.41 *Aneumastus*

Class: Bacillariophyceae
Order: Naviculales
Family: Mastogloiaceae

Diatoms

Mastogloia

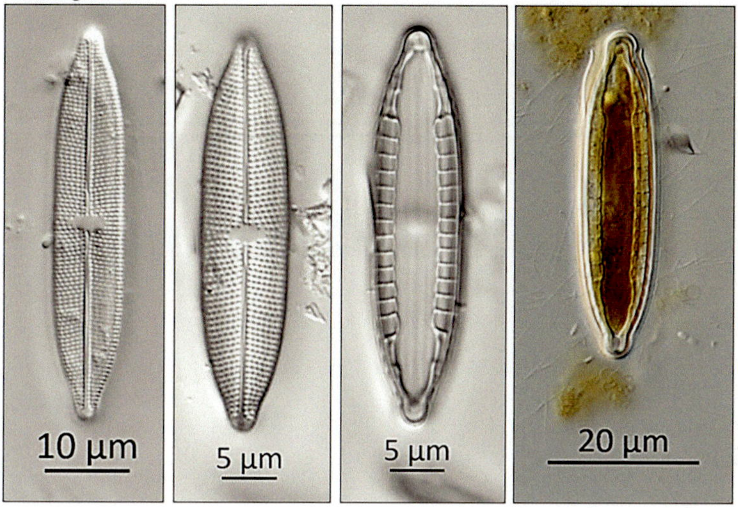

Mastogloia calcarea

Periphyton

Fig. 3.42 *Mastogloia*

Class: Bacillariophyceae
Order: Naviculales
Family: Amphipleuraceae

Diatoms

scale bar = 10 µm for all diatom images, except as noted

Amphipleura

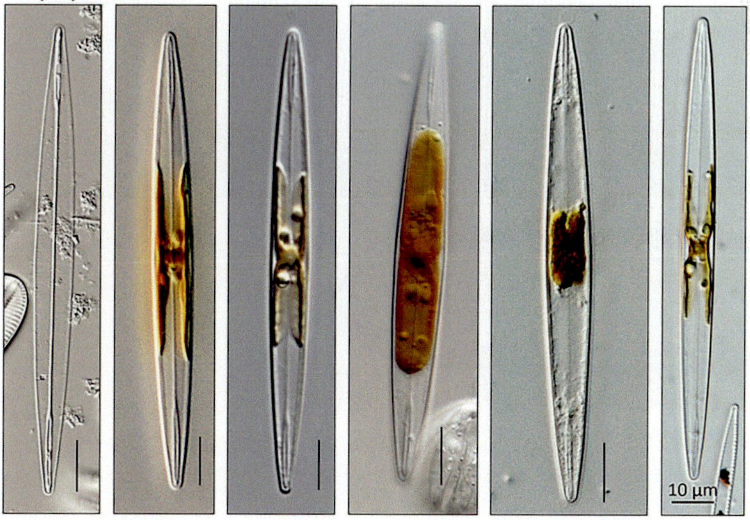

Amphipleura pellucida

Periphyton

Fig. 3.43 *Amphipleura*

Class: Bacillariophyceae
Order: Naviculales
Family: Amphipleuraceae

Diatoms

scale bar = 10 μm for all diatom images, except as noted

Frustulia

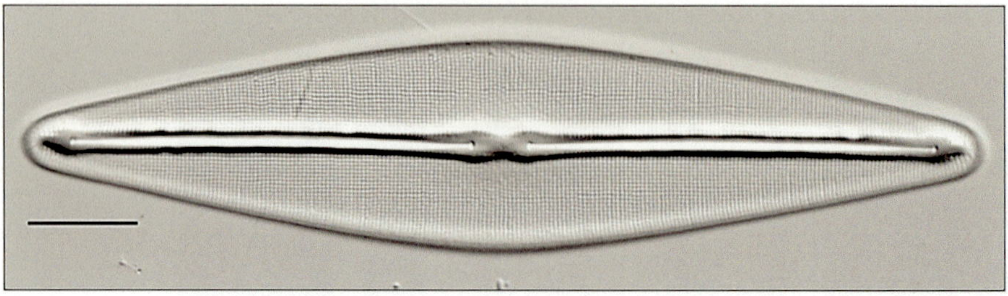

Frustulia sp. Periphyton

Fig. 3.44 *Frustulia*

Class: Bacillariophyceae
Order: Naviculales
Family: Neidiaceae

Diatoms

scale bar = 10 μm for all diatom images, except as noted

Neidium

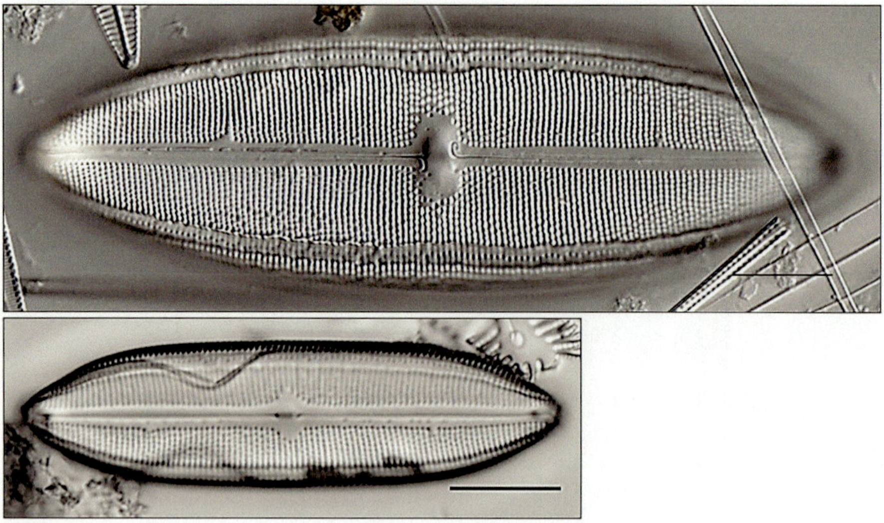

Neidium spp. Periphyton

Fig. 3.45 *Neidium*

Class: Bacillariophyceae
Order: Naviculales
Family: Naviculaceae

Diatoms

scale bar = 10 µm for all diatom images, except as noted

Hippodonta

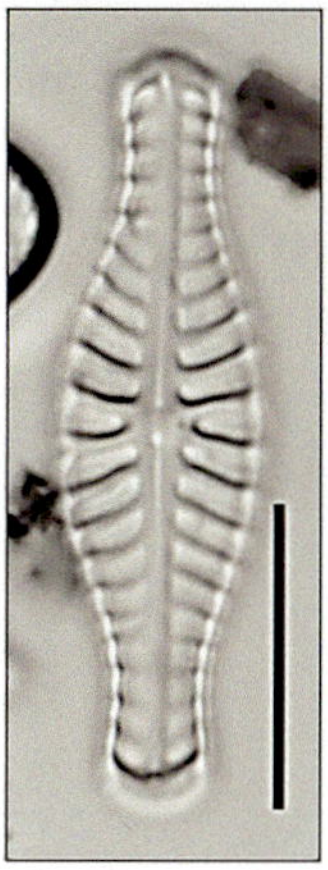

Hippodonta capitata

Periphyton

Fig. 3.46 *Hippodonta*

Class: Bacillariophyceae
Order: Naviculales
Family: Naviculaceae

Diatoms

scale bar = 10 µm for all diatom images, except as noted

Caloneis

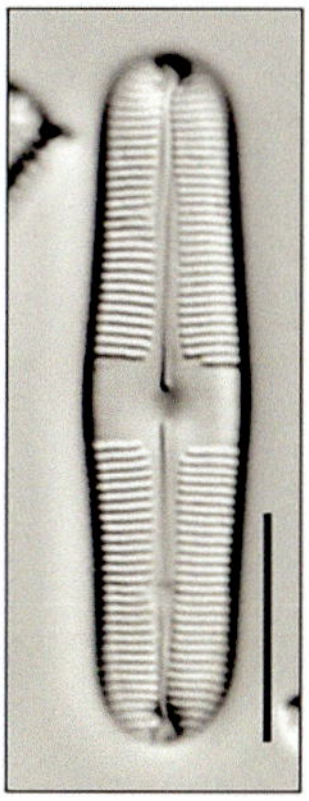

Caloneis sp.

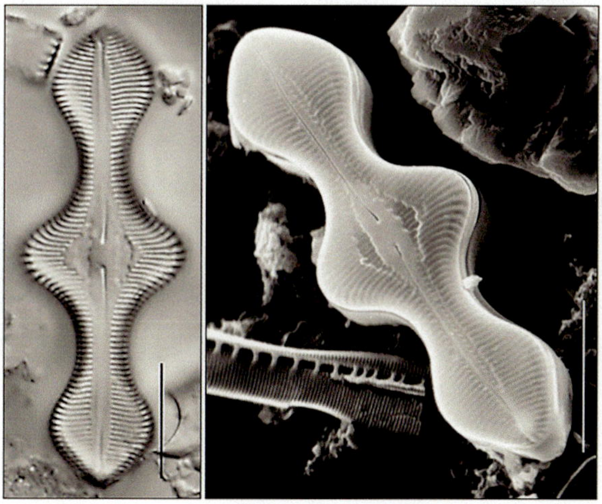

Caloneis lewisii

Periphyton

Fig. 3.47 *Caloneis*

Class: Bacillariophyceae
Order: Naviculales
Family: Diploneidaceae

Diatoms

scale bar = 10 µm for all diatom images, except as noted

Diploneis

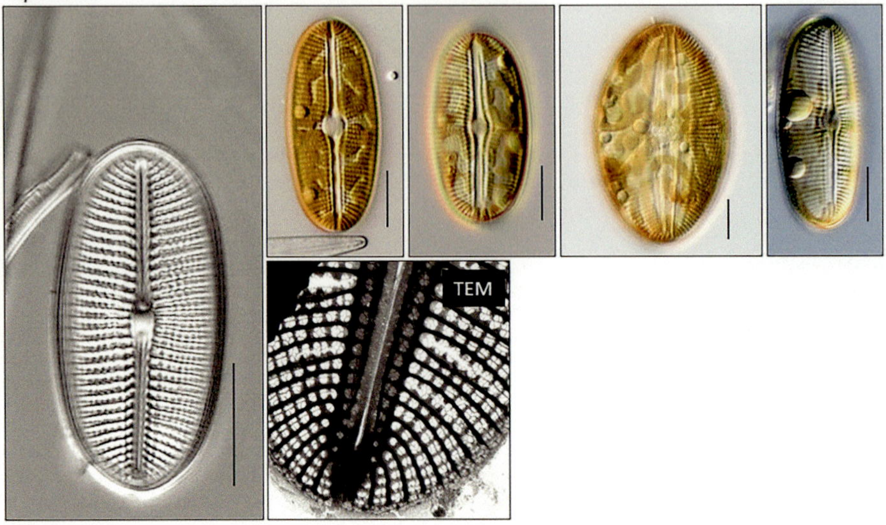

Diploneis spp.

Periphyton

Fig. 3.48 *Diploneis*

Class: Bacillariophyceae
Order: Cymbellales
Family: Cymbellaceae

Diatoms

scale bar = 10 µm for all diatom images, except as noted

Encyonema

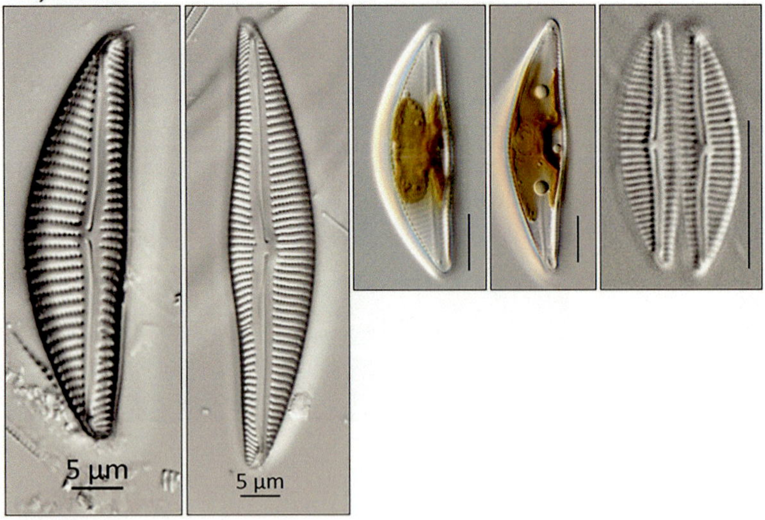

Encyonema spp.

Periphyton

Fig. 3.49 *Encyonema*

Class: Bacillariophyceae
Order: Cymbellales
Family: Cymbellaceae

Diatoms

Encyonopsis

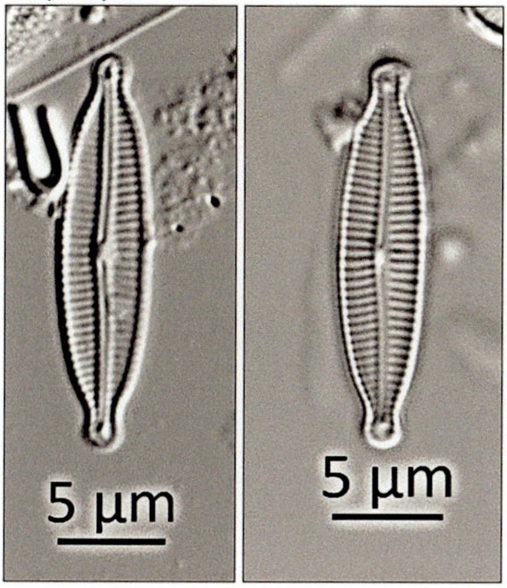

Encyonopsis sp.

Periphyton

Fig. 3.50 *Encyonopsis*

Class: Bacillariophyceae
Order: Naviculales
Family: Naviculaceae

Diatoms

Gyrosigma

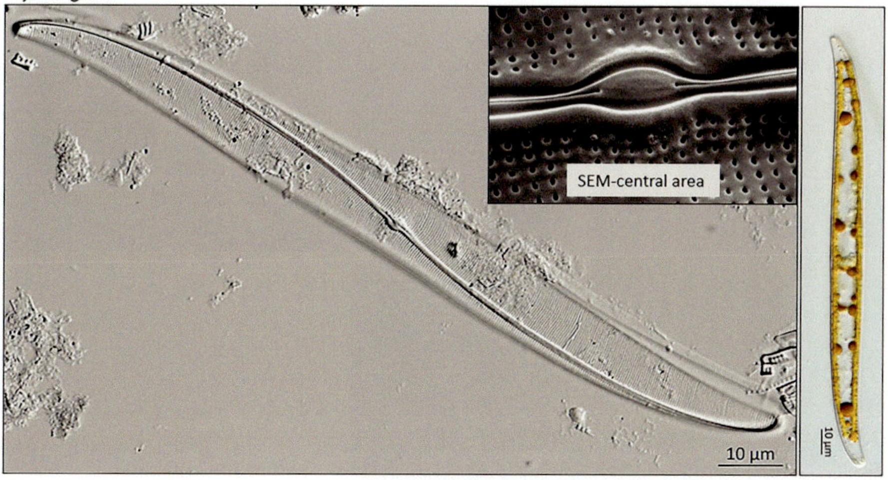

Gyrosigma obscurum

Periphyton

Fig. 3.51 *Gyrosigma*

Class: Bacillariophyceae
Order: Naviculales
Family: Plagiotropidaceae

Diatoms

Plagiotropis

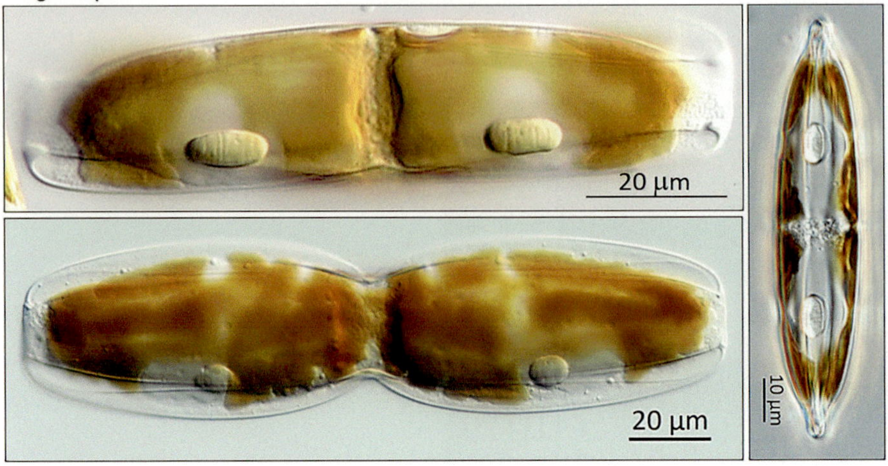

Plagiotropis sp. Periphyton

Fig. 3.52 *Plagiotropis*

Class: Bacillariophyceae
Order: Thalassiophysales
Family: Catenulaceae

Diatoms

scale bar = 10 µm for all diatom images, except as noted

Amphora

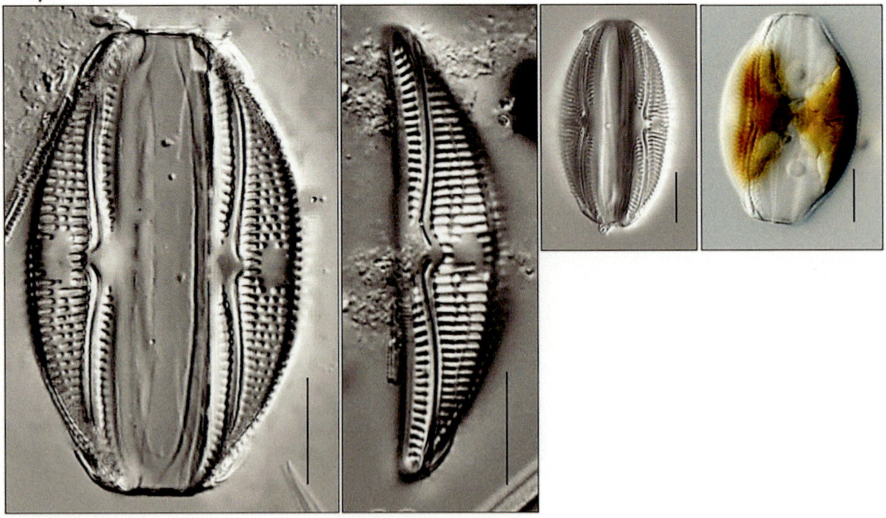

Amphora spp. Periphyton

Fig. 3.53 *Amphora*

Class: Bacillariophyceae
Order: Cymbellales
Family: Cymbellaceae

Diatoms

scale bar = 10 μm for all diatom images, except as noted

Cymbella

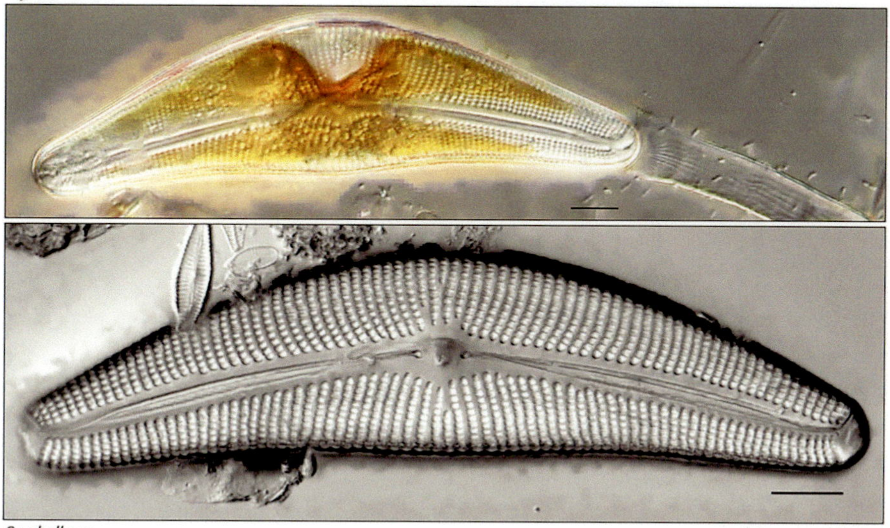

Cymbella sp.

Periphyton, epiphyton

Fig. 3.54 *Cymbella*

Class: Bacillariophyceae
Order: Cymbellales
Family: Rhoicospheniaceae

Diatoms

scale bar = 10 μm for all diatom images, except as noted

Rhoicosphenia

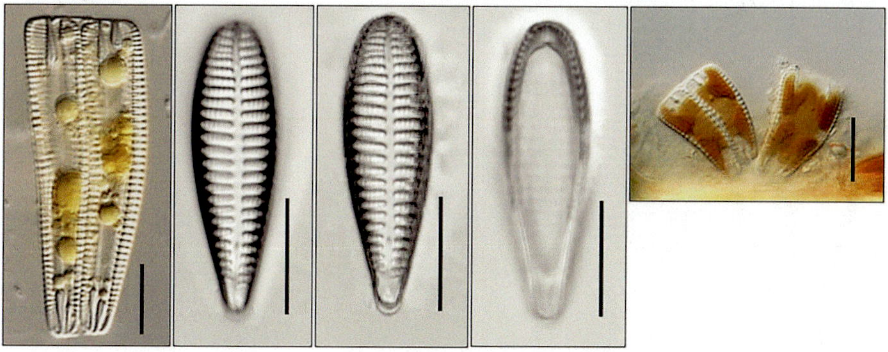

Rhoicosphenia spp.

Periphyton, epiphyton

Fig. 3.55 *Rhoicosphenia*

Class: Bacillariophyceae
Order: Cymbellales
Family: Gomphonemataceae

Diatoms

Reimeria

scale bar = 10 µm for all diatom images, except as noted

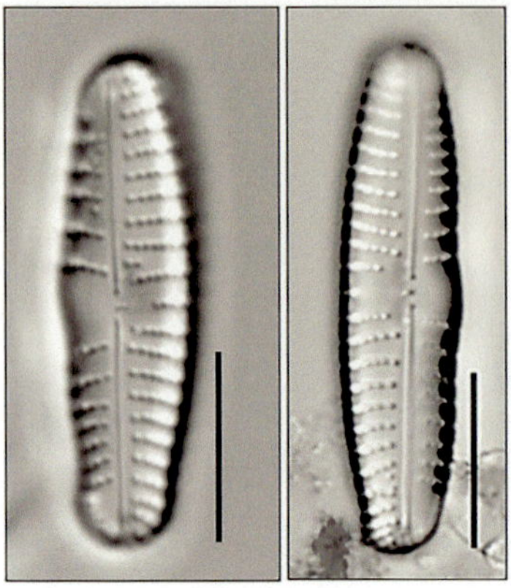

Reimeria uniseriata Periphyton

Fig. 3.56 *Reimeria*

Class: Bacillariophyceae
Order: Cymbellales
Family: Gomphonemataceae

Diatoms

Gomphonema

Valve view Girdle view

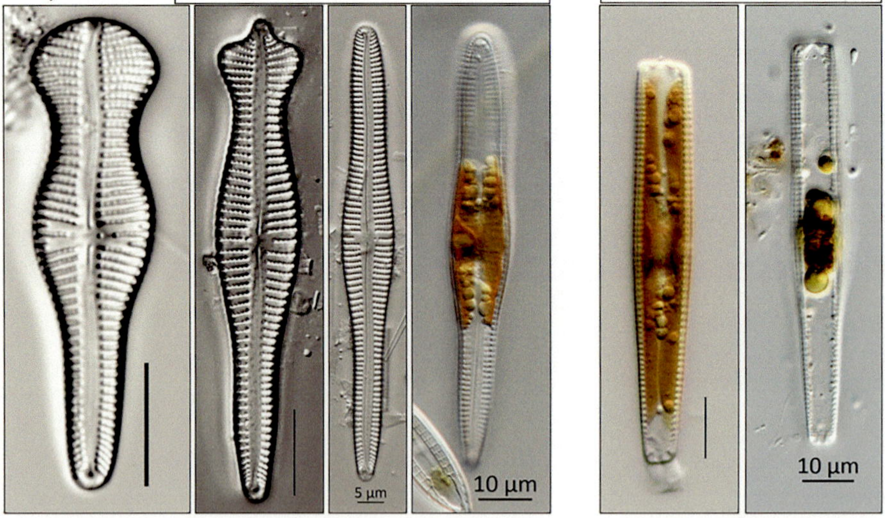

Gomphonema spp. Periphyton, epiphyton

Fig. 3.57 *Gomphonema*

Class: Bacillariophyceae
Order: Cymbellales
Family: Gomphonemataceae

Diatoms

scale bar = 10 µm for all diatom images, except as noted

Gomphoneis

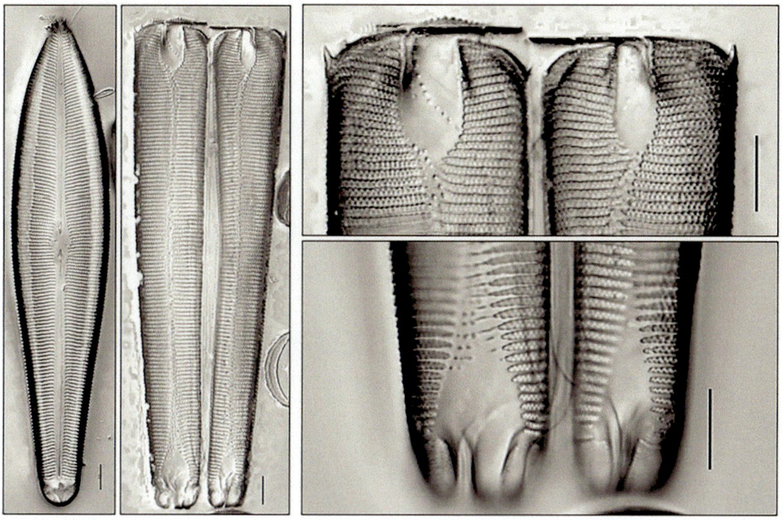

Gomphoneis mammilla

Periphyton

Fig. 3.58 *Gomphoneis*

Class: Bacillariophyceae
Order: Cymbellales
Family: Cymbellaceae

Diatoms

Didymosphenia (aka Rock Snot)

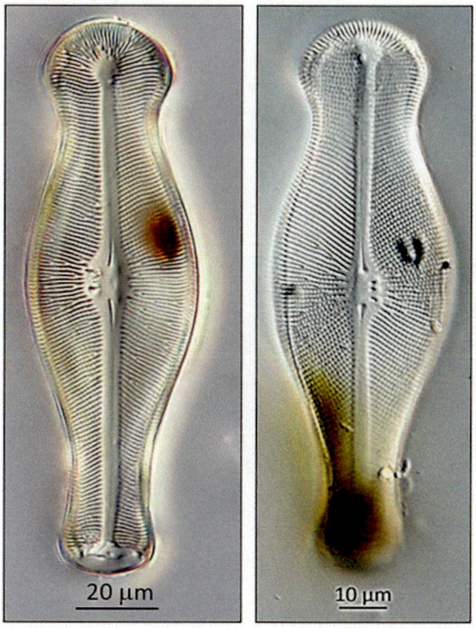

Didymosphenia geminata

Periphyton, epilithon

Fig. 3.59 *Didymosphenia (aka Rock Snot)*

Class: Bacillariophyceae
Order: Rhopalodiales
Family: Rhopalodiaceae

Diatoms

scale bar = 10 µm for all diatom images, except as noted

Epithemia

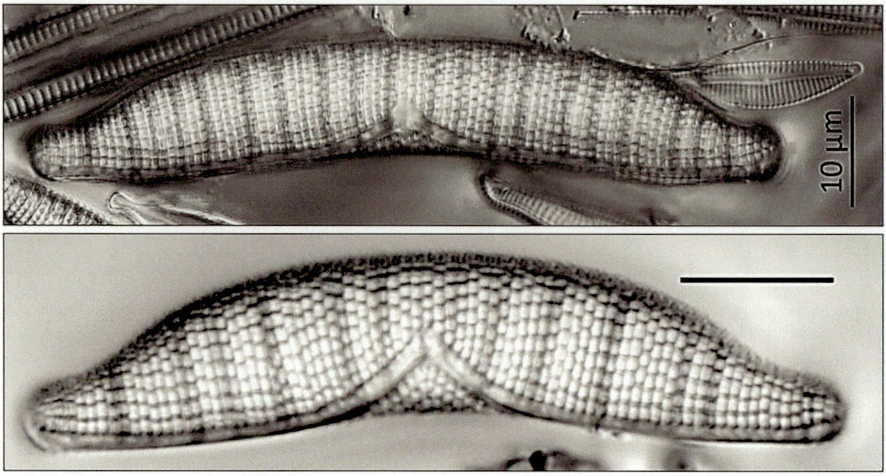

Epithemia adnanta

Epiphyton

Fig. 3.60 *Epithemia*

Class: Bacillariophyceae
Order: Rhopalodiales
Family: Rhopalodiaceae

Diatoms

scale bar = 10 µm for all diatom images, except as noted

Epithemia (formerly *Rhopalodia*)

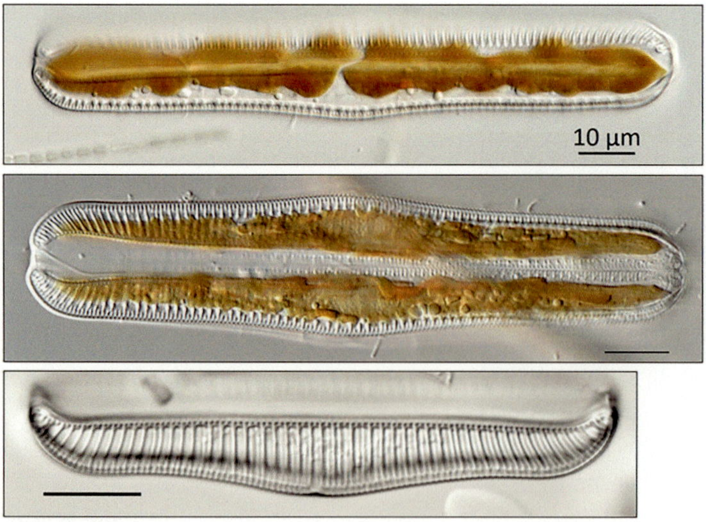

Epithemia gibba

Periphyton

Fig. 3.61 *Epithemia* (formerly *Rhopalodia*)

Class: Bacillariophyceae
Order: Bacillariales
Family: Bacillariaceae

Diatoms

Nitzschia

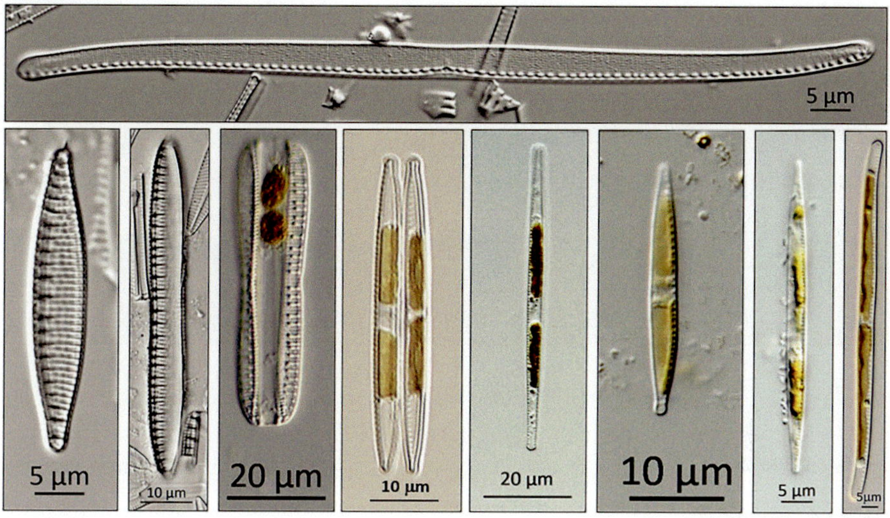

Nitzschia spp.

Periphyton, epiphyton

Fig. 3.62 *Nitzschia*

Class: Bacillariophyceae
Order: Bacillariales
Family: Bacillariaceae

Diatoms

Hantzschia

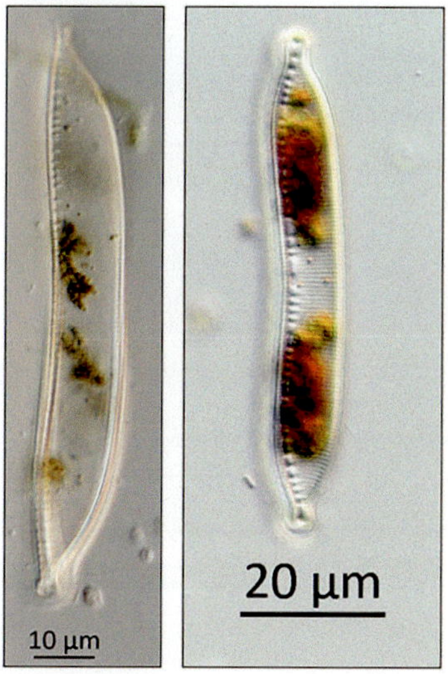

Hantzschia spp.

Periphyton, moist soils

Fig. 3.63 *Hantzschia*

Class: Bacillariophyceae
Order: Bacillariales
Family: Bacillariaceae

Diatoms

Bacillaria

Bacillaria paxillifera Periphyton, epiphyton

Fig. 3.64 *Bacillaria*

Class: Bacillariophyceae
Order: Bacillariales
Family: Bacillariaceae

Diatoms

scale bar = 10 µm for all diatom images, except as noted

Tryblionella

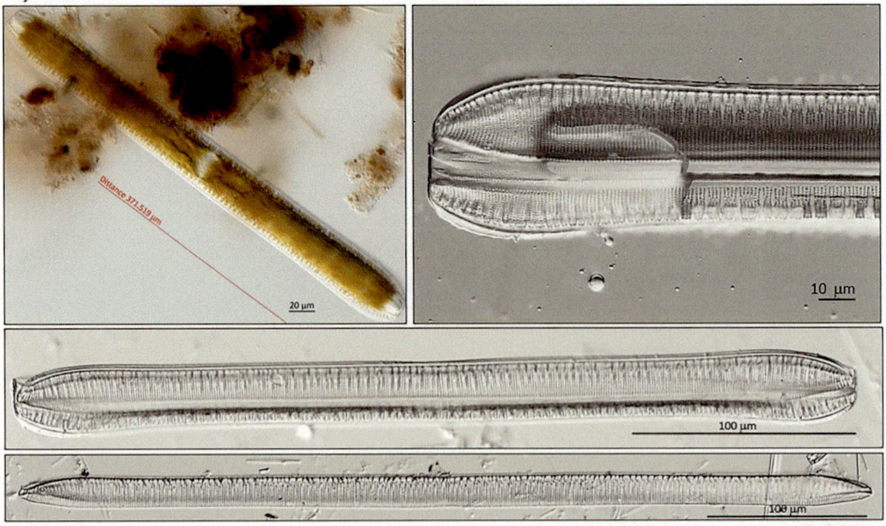

Tryblionella spp. Periphyton

Fig. 3.65 *Tryblionella*

Class: Bacillariophyceae
Order: Bacillariales
Family: Bacillariaceae

Diatoms

scale bar = 10 μm for all diatom images, except as noted

Tryblionella

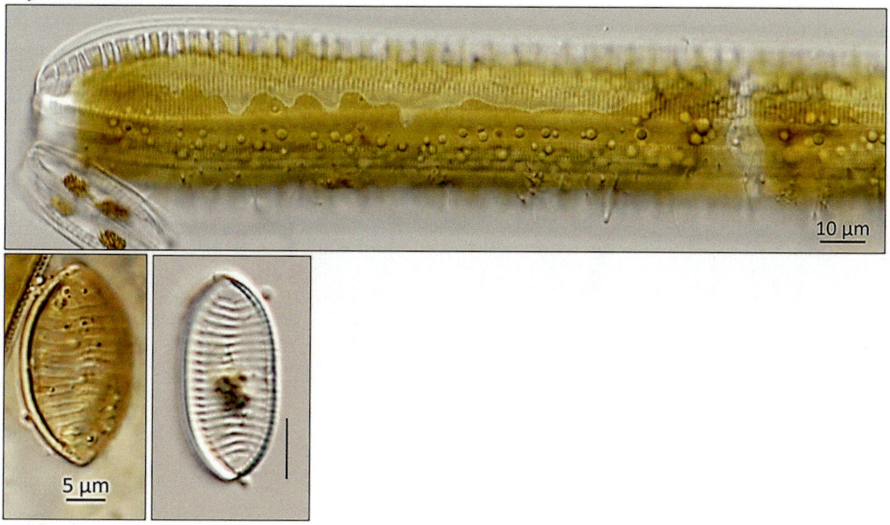

Tryblionella spp.

Periphyton

Fig. 3.66 *Tryblionella*

Class: Bacillariophyceae
Order: Bacillariales
Family: Bacillariaceae

Diatoms

scale bar = 10 μm for all diatom images, except as noted

Psammodictyon

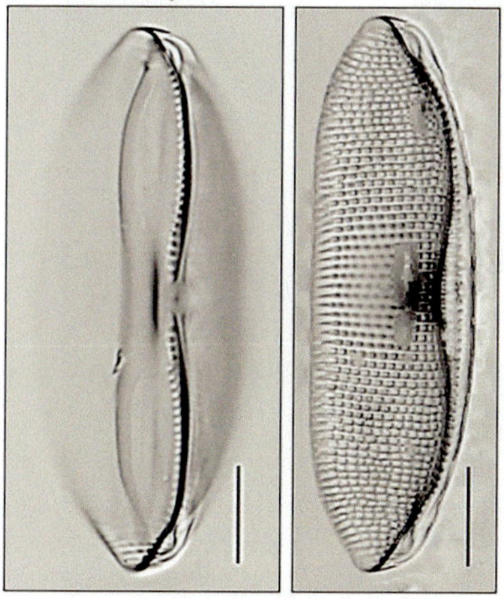

Psammodictyon constrictum

Periphyton

Fig. 3.67 *Psammodictyon*

Class: Bacillariophyceae
Order: Bacillariales
Family: Bacillariaceae

Diatoms

scale bar = 10 µm for all diatom images, except as noted

Grunowia

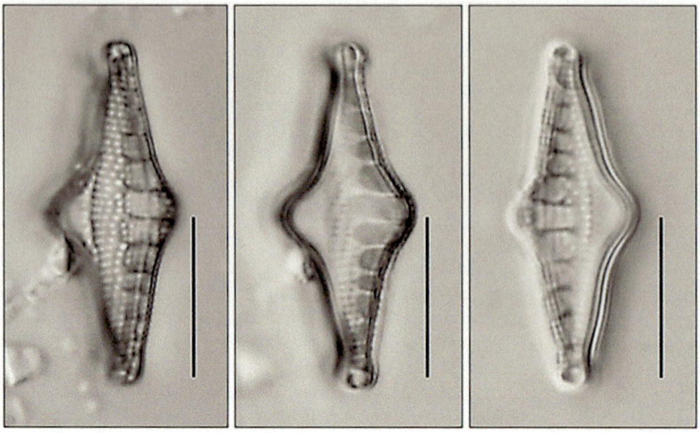

Grunowia tabellaria

Periphyton

Fig. 3.68 *Grunowia*

Class: Bacillariophyceae
Order: Surirellales
Family: Surirellaceae

Diatoms

scale bar = 10 µm for all diatom images, except as noted

Surirella (formerly *Cymatopleura*)

Surirella undulata

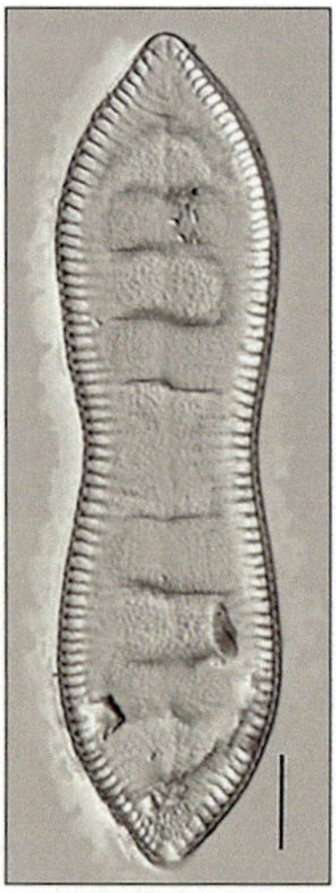

Surirella librile

Periphyton

Fig. 3.69 *Surirella* (formerly *Cymatopleura*)

Class: Bacillariophyceae
Order: Surirellales
Family: Surirellaceae

Diatoms

scale bar = 10 µm for all diatom images, except as noted

Iconella (formerly *Surirella*)

Surirella spp.

Periphyton

Fig. 3.70 *Iconella* (formerly *Surirella*)

Class: Bacillariophyceae
Order: Surirellales
Family: Surirellaceae

Diatoms

Iconella (formerly *Campylodiscus*)

Iconella hibernica

Periphyton

Fig. 3.71 *Iconella* (formerly *Campylodiscus*)

Class: Bacillariophyceae
Order: Surirellales
Family: Surirellaceae

Diatoms

Iconella (formerly *Campylodiscus*)

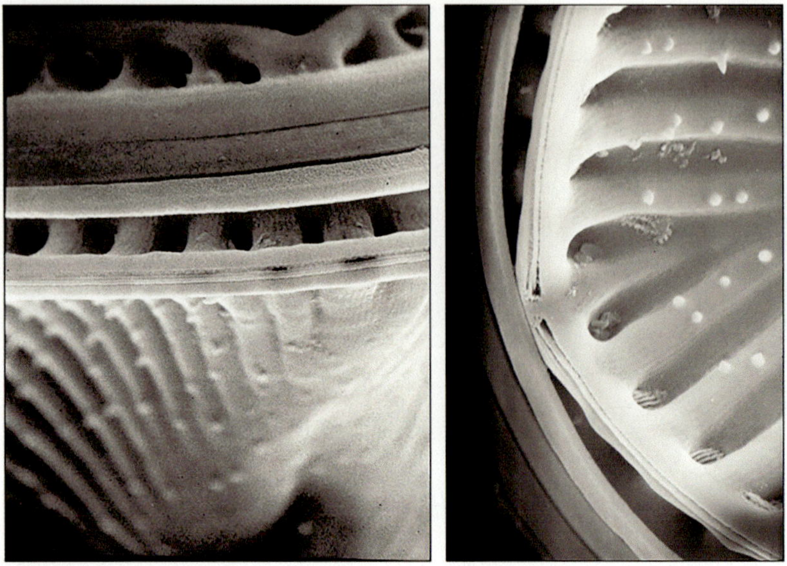

Fig. 3.72 *Iconella* (formerly *Campylodiscus*)

Class: Bacillariophyceae
Order: Surirellales
Family: Entomoneidaceae

Diatoms

scale bar = 10 μm for all diatom images, except as noted

Entomoneis

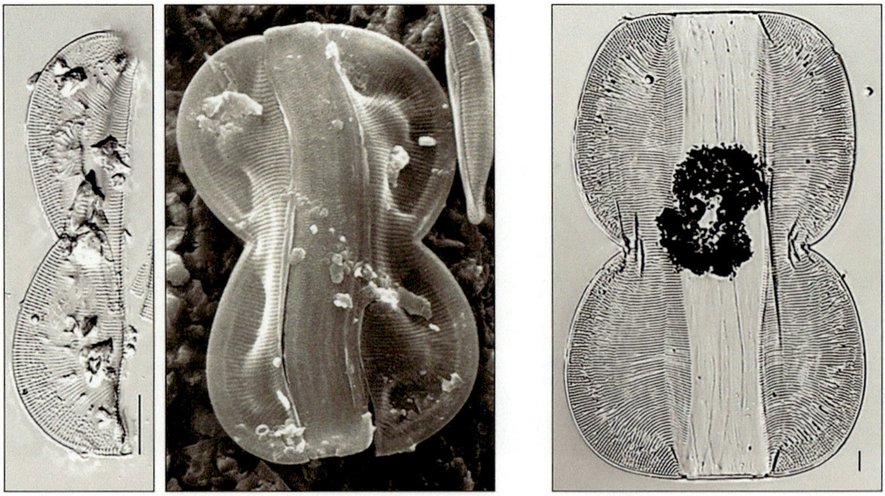

Fig. 3.73 *Entomoneis*

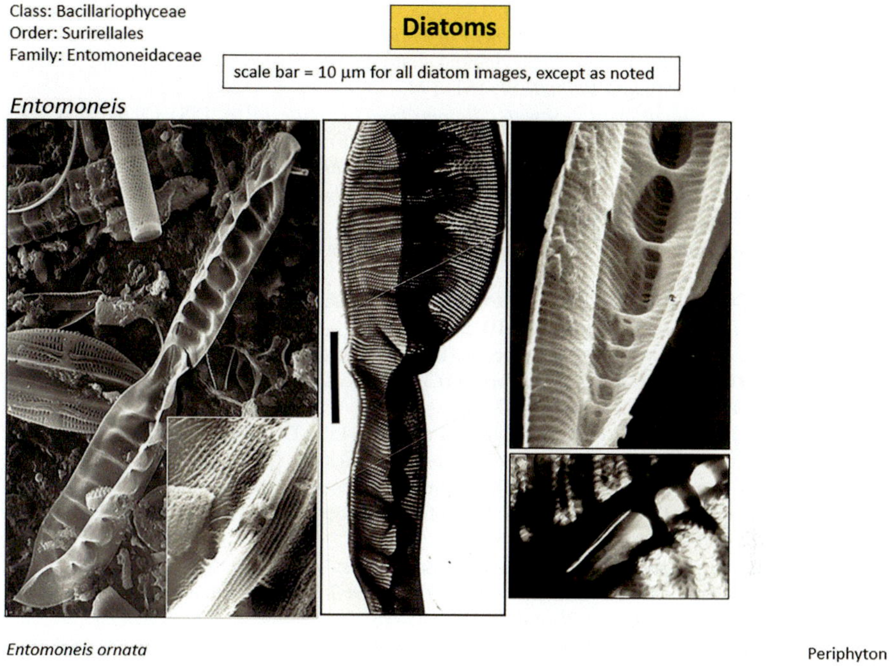

Fig. 3.74 *Entomoneis*

Suggested Reading

Abbriano, R. (2023). Gliding toward new discoveries in diatom adhesion and motility (algae highlight). *Journal of Phycology, 59*(1), 52–53.

Academy of Natural Sciences. (2014). NADED Master-list of names and taxa codes for diatoms and other algae: Names used in the North American Diatom Ecological Database. Phycology Section, Patrick Center for Environmental Research, The Academy of Natural Sciences, Philadelphia, PA. Retrieved 5 December 2014 from http://diatom.ansp.org/dpdc/taxalist.asp

Anonymous. (1975). Proposals for a standardization of diatom terminology and diagnoses. *Nova Hedwigia Beiheft, 53*, 323–365.

Ashworth, M. P., Lobban, C. S., Witkowski, A., Theriot, E. C., Sabir, M. J., Baeshen, M. N., Hajarah, N. H., Baeshen, N. A., Sabir, J. S., & Jansen, R. K. (2016). Molecular and morphological investigations of the stauros-bearing, raphid pennate diatoms (Bacillariophyceae): *Craspedostauros* E.J. Cox, and *Staurotropis* T.B.B. Paddock, and their relationship to the rest of the Mastogloiales. *Protist, 168*(1), 48–70.

Bahls, L. (2013). New diatoms (Bacillariophyta) from western North America. *Phytotaxa, 82*(1), 7–28.

Bahls, L. (2019). *Diatoms from Western North America 2. Cymbellafalsa, Cymbopleura and Delicatophycus (Bacillariophyta) — Taxonomy, ecology, biogeography* (pp. [1]–114).

Bahls, L. (2021). Diatoms of Montana and western North America: Catalog and atlas of species in the Montana diatom collection, volume 1. *Academy of Natural Sciences of Philadelphia Special Publication, 24*, 1–508.

Bothwell, M. L., Suzuki, K. E., Bolin, M. K., & Hardy, F. J. (1989). Evidence of dark avoidance by phototrophic periphytic diatoms in lotic systems. *Journal of Phycology, 25*, 85–94.

Camburn, K. E., & Charles, J. C. (2000). Diatoms of low-alkalinity lakes in the northeastern United States. *Academy of Natural Sciences of Philadelphia, Special Publication, 18*, 1–152.
Cox, E. J. (1977). Raphe structure in naviculoid diatoms as revealed by the scanning electron microscope. *Nova Hedwigia Beiheft, 54*, 161–174.
Cox, E. J. (1996). *Identification of freshwater diatoms from live material* (p. 158). Chapman & Hall.
Cox, E. J. (2015). Coscinodiscophyceae, Mediophyceae, Fragilariophyceae, Bacillariophyceae (Diatoms). In W. Frey (Ed.), *Syllabus of plant families. Adolf Engler's syllabus der Pflanzenfamilien. 13th Ed. Photoautotrophic eukaryotic algae Glaucocystophyta, Cryptophyta, Dinophyta/Dinozoa, Heterokontophyta/Ochrophyta, Chlorarachniophyta/Cercozoa, Euglenophyta/Euglenozoa, Chlorophyta, Streptophyta p.p* (pp. 64–103). Borntraeger Science Publishers.
Hustedt, F. (1930). Bacillariophyta (Diatomeae) Zweite Auflage. In A. Pascher (Ed.), *Die Süsswasser-Flora Mitteleuropas* (Heft 10. pp. [i]–viii, [1]–466). Verlag von Gustav Fischer.
Johansen, J.[R.], Kociolek, [J.]P. & Lowe, R.[L.]. (2008). *Spicaticribra kingstonii*, gen. nov. et sp. nov. (Thalassiosirales, Bacillariophyta) from Great Smoky Mountains National Park, U.S.A. *Diatom Research, 23*(2), 367–375.
Kingston, J. C. (2003). Araphid and Monoraphid diatoms. In J. D. Wehr & R. G. Sheath (Eds.), *Freshwater algae of North America, ecology and classification* (pp. 595–636). Academic.
Kociolek, J. P., & Kingston, J. C. (1999). Taxonomy, ultrastructure and distribution of some gomphonemoid diatoms (Bacillariophyceae: Gomphonemataceae) from rivers in the United States. *Canadian Journal of Botany, 77*(5), 686–705.
Kociolek, J. P., & Rosen, B. H. (1984). Observations on North American *Gomphoneis* (Bacillariophyceae). I. Valve ultrastructure of *G. mammilla* with comment on the taxonomic status of the genus. *Journal of Phycology, 20*, 361–368.
Kociolek, J. P., & Stoermer, E. F. (1988). Taxonomy, ultrastructure and distribution of *Gomphoneis herculeana, G. eriense* and closely related species (Naviculales: Gomphonemataceae). *Proceedings of the Academy of Natural Sciences of Philadelphia, 140*(2), 24–97.
Kociolek, F. P., Stoermer, A. F., & Edlund, M. A. (1995). Two new freshwater diatom species. In J. P. Kociolek & M. J. Sullivan (Eds.), *A century of diatom research in North America: A tribute to the distinguished carrers of Charles W. Reimer and Ruth Patrick* (pp. 9–19).
Kulikovskiy, M., Lange-Bertalot, H., Annenkova, N., Gusev, E., & Kociolek, J. P. (2016). Morphological and molecular evidence support description of two new diatom species from the genus *Ulnaria* in Lake Baikal. *Fottea, 16*(1), 34–42.
Lowe, R. L. (1975). Comparative ultrastructure of the valves of some *Cyclotella* species (Bacillariophyceae). *Journal of Phycology, 11*, 415–424.
Lowe, R. L., & Kociolek, J. P. (1984). New and rare diatoms from Great Smoky Mountains National Park. *Nova Hedwigia, 39*(3–4), 465–476.
Patrick, R. [M.] & Reimer, C. W. (1966). The diatoms of the United States exclusive of Alaska and Hawaii. Volume 1: Fragilariaceae, Eunotiaceae, Achnanthaceae, Naviculaceae. *Monographs of the Academy of Natural Sciences of Philadelphia, 13*, 1–688.
Patrick, R., & Reimer, C. W. (1975). *The diatoms of the United States exclusive of Alaska and Hawaii* (Vol. 2, Part 1. Entomoneidaceae, Cymbellaceae, Gomphonemaceae, Epithemiaceae. pp. [i]–ix, [1]–213, 28 pls [in text]). The Academy of Natural Sciences of Philadelphia.
Spaulding, et al. (2021). Diatoms.org: Supporting taxonomists, connecting communities. *Diatom Research, 36*(4), 291–304. https://doi.org/10.1080/0269249X.2021.2006790

Chrysophytes and Haptophytes

4

Chrysophytes are a group of microscopic unicellular or colonial eukaryotic algae. Once classified with the diatoms and brown algae, they are also within the kingdom Chromista, phylum heterokontophyta. Chrysophytes have two flagella that are unequal in length. On the short flagellum, there is a photoreceptor, and they also have a stigma (eyespot) in one of the chromoplasts. Many Chrysophytes are single-cell organisms that swim independently, while others form colonies that tumble through the water and may superficially resemble the motile colonies of green algae. Some are found attached to various substrata, and no motile stage has been observed.

Chrysophytes have chlorophyll *a* and *c* and are commonly called the golden-brown algae. Like the diatoms, they also have fucoxanthin that gives them their brownish color, as well as other carotenoids. The main storage compound from photosynthesis is chrysolaminarin, once known as leucosin, with beta-beta linked glucose units. Many have no pigments, thus are not photosynthetic, and can either absorb organic compounds or ingest particles and bacteria by phagocytosis. Because of this, some would categorize these organisms as protists, a term that broadly includes many of the single-celled algae. Some have both photosynthetic capabilities and are phagocytotic, such as *Chrysamoeba* (*see* Fig. 4.6).

Some species make silica scales that are produced inside the cells and then deposited outside the cell membrane. The scales may also have spines, which create an impressive armor against zooplankton. The unique structure of the scales and spines, along with a silicified resting stage, stomotocysts, is used for taxonomic identification. While morphological characteristics are traditionally used to identify the organisms in the Chrysophytes, the relationship between groups using molecular tools has realigned some of the thinking about nearest relatives.

Chrysophytes are divided into several orders with members that are mostly found in freshwater environments. Some species prefer colder water, with higher populations found in the winter, and some have a specific pH range they prefer. One genus, *Synura* (*see* Fig. 4.19), is known to impart a foul fishy odor that can be problematic for the quality of drinking water; however, since it is not known to be toxic, this is considered a nuisance rather than a harmful organism. The Phaeothamniophyceae, a closely related class, are included in this section as well as *Prymnesium* (*see* Fig. 4.23) in the phylum Haptophyta, a brackish water relative that is known to be toxic to fish (Figs. 4.1 to 4.23).

B. H. Rosen, *Color Atlas of Freshwater Algae*,
https://doi.org/10.1007/978-3-031-93122-2_4

Class: Chrysophyceae
Order: Chromulinales
Family: Chromulinaceae

Chrysophytes

Chromulina

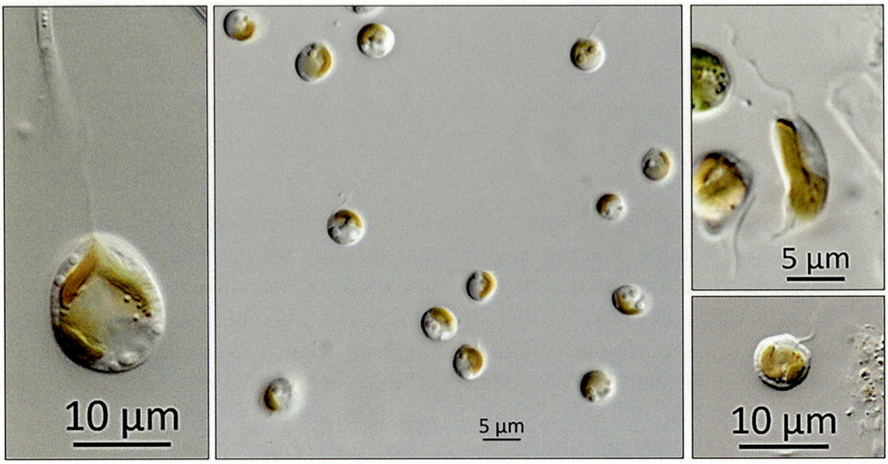

Chromulina spp.

Plankton

Fig. 4.1 *Chromulina*

Class: Chrysophyceae
Order: Ochromonadales
Family: Ochromonadaceae

Chrysophytes

Ochromonas

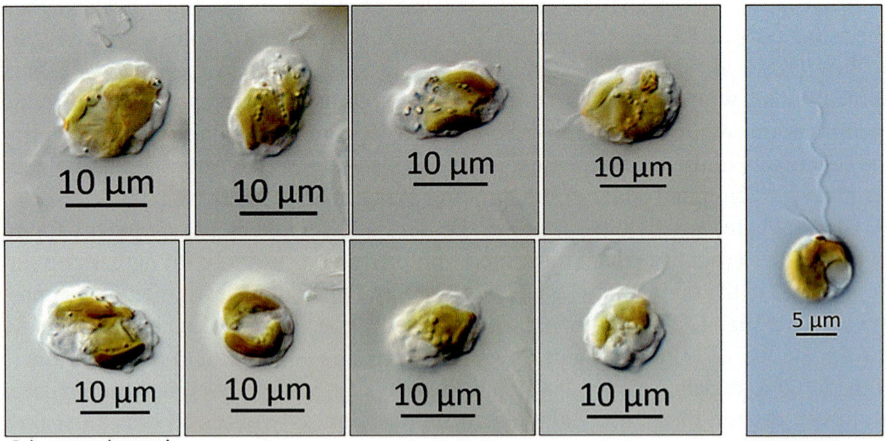

Ochromonas hovassei
[species taxonomic status uncertain]

Ochromonas sp.

Plankton, metaphyton

Fig. 4.2 *Ochromonas*

Class: Chrysophyceae
Order: Ochromonadales
Family: Ochromonadaceae

Chrysophytes

Ochromonas

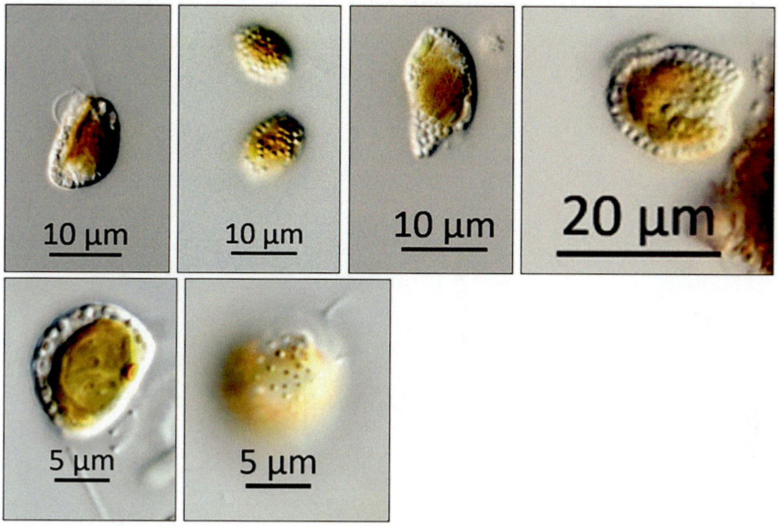

Ochromonas allorgei

Plankton, metaphyton

Fig. 4.3 *Ochromonas*

Class: Chrysophyceae
Order: Ochromonadales
Family: Ochromonadaceae

Chrysophytes

Ochromonas

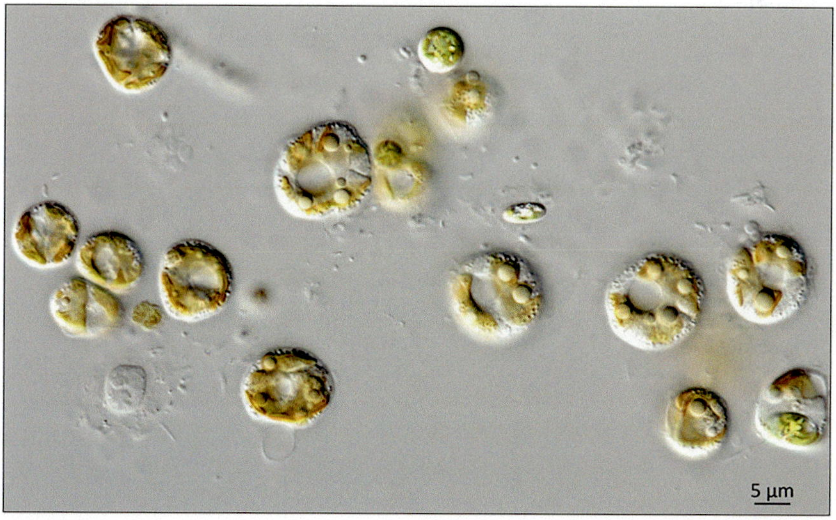

Ochromonas sp.

Plankton, metaphyton

Fig. 4.4 *Ochromonas*

Class: Chrysophyceae
Order: Chrysosaccales
Family: Stylococcaceae

Chrysophytes

Lagynion

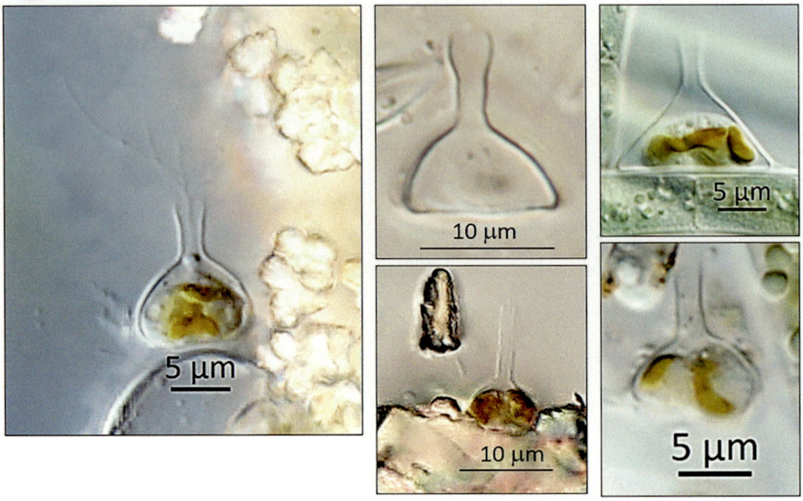

Lagynion spp. Epiphyton

Fig. 4.5 *Lagynion*

Class: Chrysophyceae
Order: Chromulinales
Family: Chrysamoebaceae

Chrysophytes

Chrysamoeba

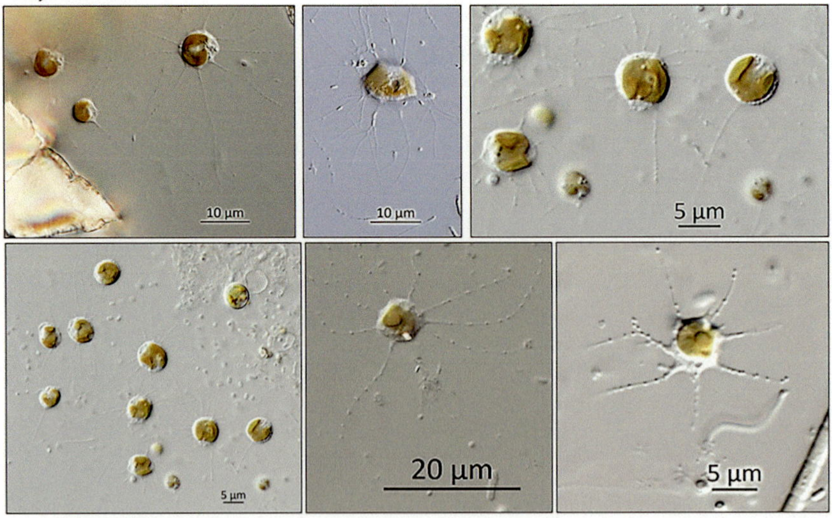

Chrysamoeba radians Epiphyton

Fig. 4.6 *Chrysamoeba*

Class: Chrysophyceae
Order: Chrysosphaerales
Family: Chrysosphaeraceae

Chrysophytes

Chrysosphaera

Chrysosphaera gallica

Epiphyton

Fig. 4.7 *Chrysosphaera*

Class: Chrysophyceae
Order: Chromulinales
Family: Chrysocapsaceae

Chrysophytes

Gloeochrysis

Gloeochrysis turfosa

Naegeliella

Naegeliella flagellifera

Epiphyton

Fig. 4.8 *Gloeochrysis*

Class: Chrysophyceae
Order: Chromulinales
Family: Chrysocapsaceae

Chrysophytes

Chrysocapsa

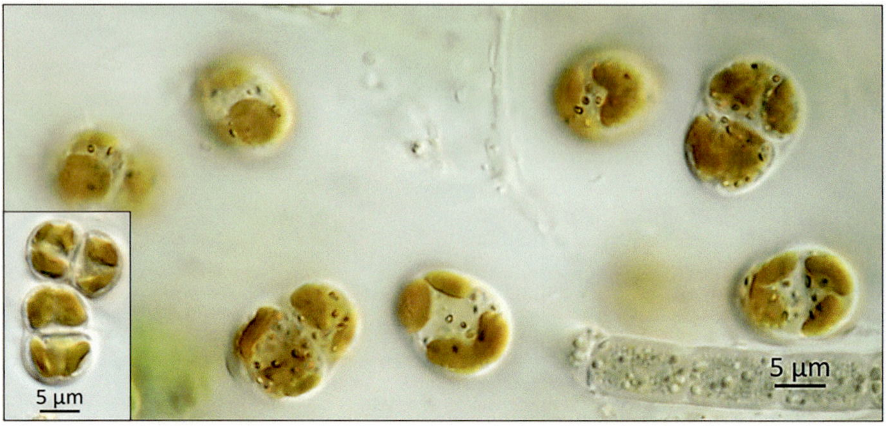

Chrysocapsa sordida Epiphyton

Fig. 4.9 *Chrysocapsa*

Class: Chrysophyceae
Order: Chrysosaccales
Family: Chrysosaccaceae

Chrysophytes

Chrysosaccus

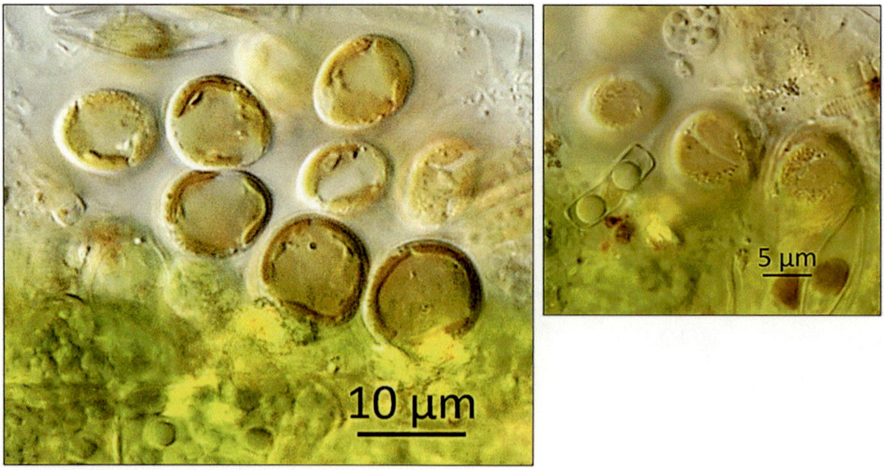

Chrysosaccus epilithicus Epiphyton

Fig. 4.10 *Chrysosaccus*

Class: Chrysophyceae
Order: Chromulinales
Family: Chromulinaceae

Chrysophytes

Chrysoxys

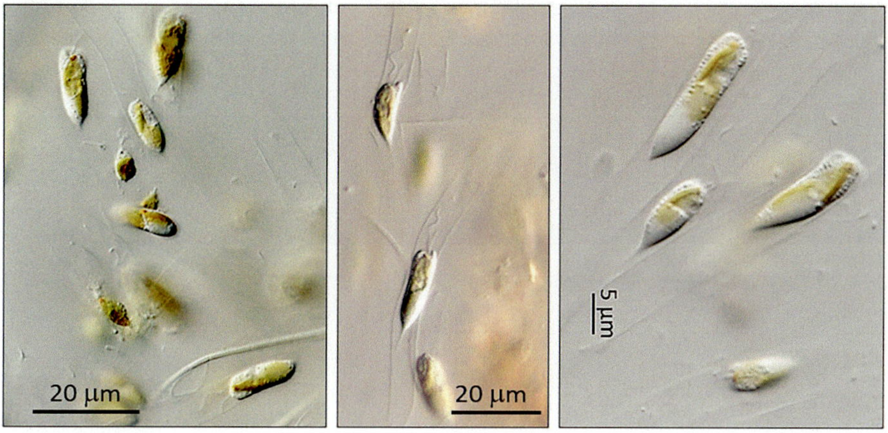

Chrysoxy maior

Epiphyton

Fig. 4.11 *Chrysoxys*

Class: Chrysophyceae
Order: Chromulinales
Family: Dinobryaceae

Chrysophytes

Dinobryon

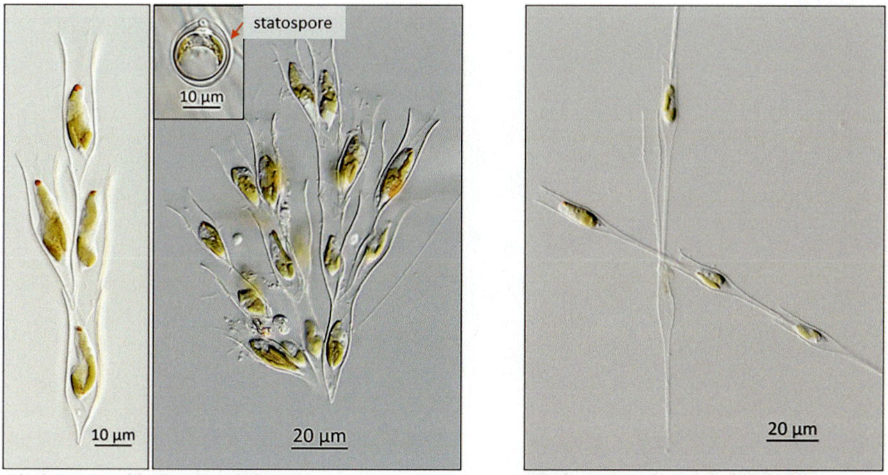

Dinobryon sertulria Ehrenberg 1834: 280

Dinobryon bavaricum Imhof 1890: 484

Phytoplankton

Fig. 4.12 *Dinobryon*

Class: Chrysophyceae
Order: Chromulinales
Family: Dinobryaceae

Chrysophytes

Epipyxis

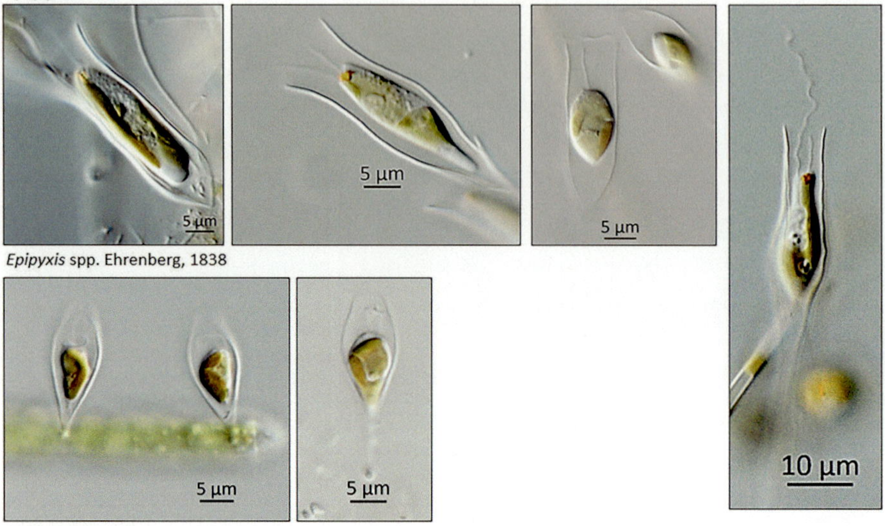

Epipyxis spp. Ehrenberg, 1838

Epipyxis tabellariae

Epiphyton

Fig. 4.13 *Epipyxis*

Class: Chrysophyceae
Order: Ochromonadales
Family: Ochromonadaceae

Chrysophytes

Uroglena

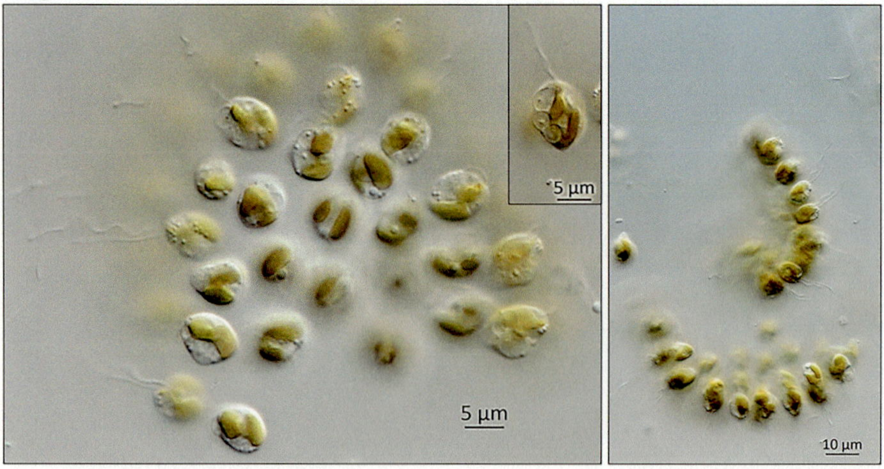

Uroglena proxima

Plankton

Fig. 4.14 *Uroglena*

Class: Chrysophyceae
Order: Phaeoplacales
Family: Chrysothallaceae

Chrysophytes

Phaeoplaca

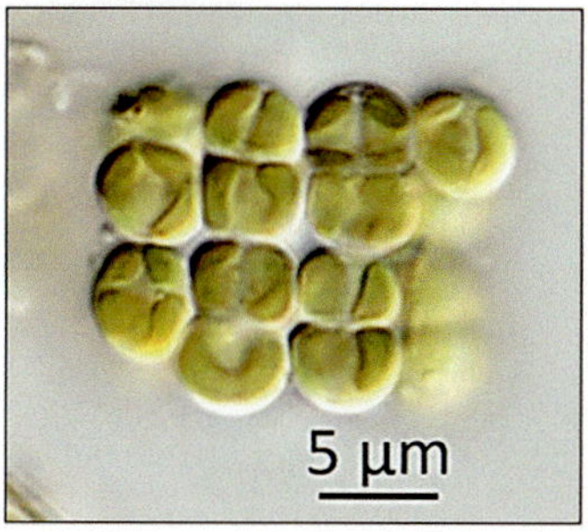

Phaeoplaca thallosa Epiphyton

Fig. 4.15 *Phaeoplaca*

Class: Chrysophyceae
Order: Hibberdiales
Family: Derepyxidaceae

Chrysophytes

Derepyxis

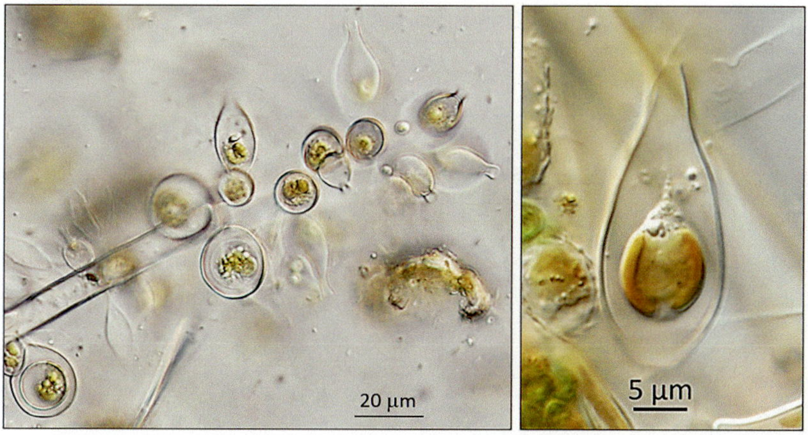

Derepyxis amphora Epiphyton

Fig. 4.16 *Derepyxis*

Class: Synurophyceae
Order: Synurales
Family: Mallomonadaceae

Chrysophytes

Mallomonas

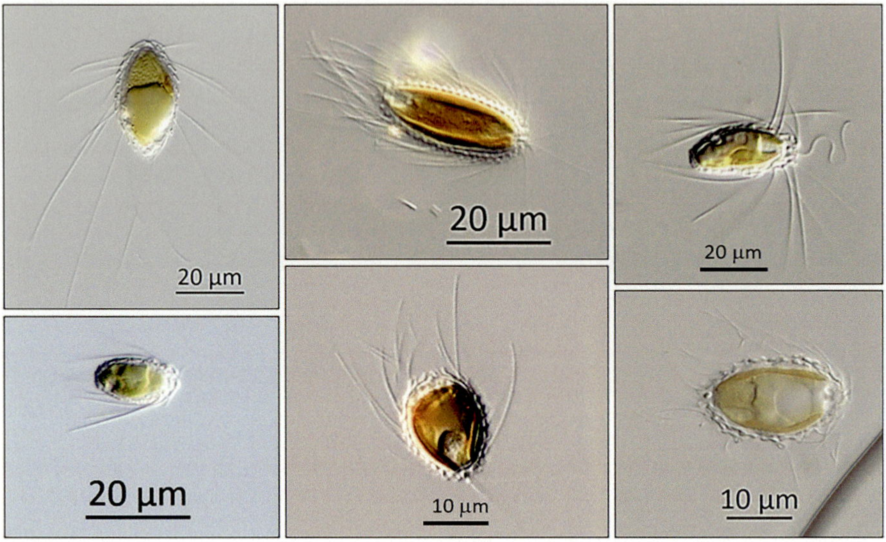

Mallomonas spp.

Fig. 4.17 *Mallomonas*

Class: Chrysophyceae
Order: Chrysophyceae *ordo incercertae sedis*
Family: Chrysophyceae *familia incertae sedis*

Chrysophytes

Randersenia

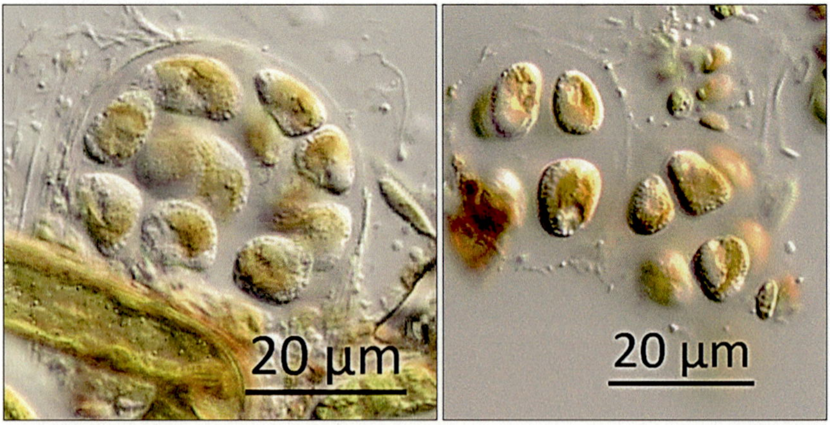

Randersenia wujekii Epiphyton

Fig. 4.18 *Randersenia*

Class: Synurophyceae
Order: Synurales
Family: Synuraceae

Chrysophytes

Synura

Synura sp. Plankton

Fig. 4.19 *Synura*

Class: Phaeothamniophyceae
Order: Phaeothamniales
Family: Phaeothamniaceae

Phaeothamniophytes

Chrysapion

Chrysapion asymmetricum Epiphyton, plankton

Fig. 4.20 *Chrysapion*

Class: Phaeothamniophyceae
Order: Phaeothamniales
Family: Phaeothamniaceae

Phaeothamniophytes

Tetrapion

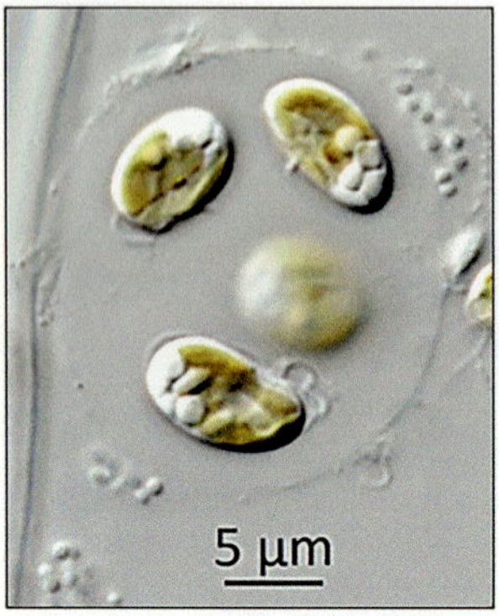

Tetrapion gloeocystiforme Epiphyton

Fig. 4.21 *Tetrapion*

Class: Phaeothamniophyceae
Order: Phaeothamniales
Family: Phaeothamniaceae

Phaeothamniophytes

Phaeoschizochlamys

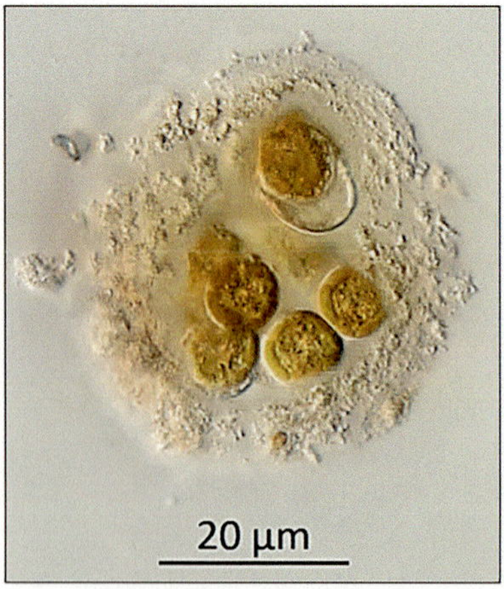

Phaeoschizochlamys delicatula Epiphyton

Fig. 4.22 *Phaeoschizochlamys*

Class: Coccolithophyceae
Order: Prymnesiales
Family: Prymnesiaceae

Haptophyta

Prymnesium

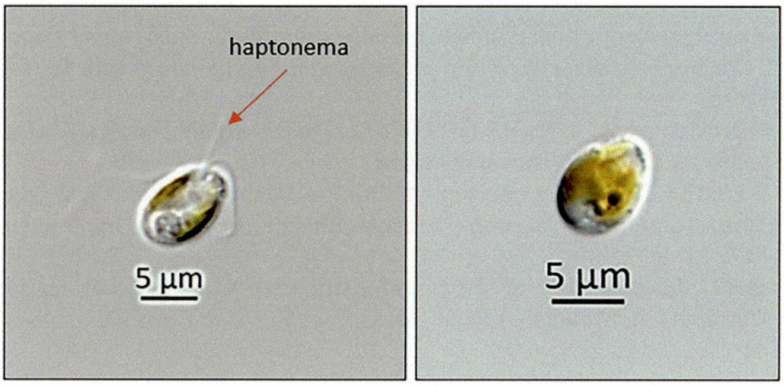

Prymnesium sp. Plankton

Fig. 4.23 *Prymnesium*

References

Chodat, R., Raineri, R., & Drew, K. (1926). Algues de la region du Grand St. Bernard III. *Bulletin de la Société Botanique de Genève, série II, 17*, 202–217.

Suggested Reading

Bock, C., Chatzinotas, A., & Boenigk, J. (2017). Genetic diversity in chrysophytes: Comparison of different gene markers. *Fottea, 17*(2), 209–221.

Bourrelly, P. (1963). Quelques Chrysophytes d'eau douce rares our nouvelles. *Protoplasma, 53*, 137–143.

Bourrelly, P. (1968). *Les algues d'eau douce. Initiation à la systématique. Tome II: Les Algues jaunes et brunes Chrysophycées, Phéophycés, Xanthophycées et Diatomées* (pp. [1–11] 12–399).

Certnerová, D., Certner, M., & Skaloud, P. (2019). Molecular phylogeny and evolution of phenotype in silica-scaled chrysophyte genus *Mallomonas*. *Journal of Phycology, 55*(4), 912–923.

Dixit, S. S., Dixit, A. S., & Smol, J. P. (1999). Lake sediment chrysophyte scales from the northeastern U.S.A. and their relationship to environmental variables. *Journal of Phycology, 35*, 903–918.

Kristiansen, J., & Preisig, H. R. (2007). Chrysophyte and Haptophyte algae. Part 2: Synurophyceae. In B. Büdel, G. Gärtner, L. Krienitz, H. R. Preisig, & M. Schagerl (Eds.), *Süßwasserflora von Mitteleuropa. Freshwater Flora of Central Europe* (Vol. 1, pp. [i–v], vi [vi–vii], 1–252). Spektrum Akademischer Verlag.

Kristiansen, J. (2005). *Golden algae. A biology of chrysophytes* (pp. iii + 1–167). A.R.G. Gantner Verlag K.G.

Nicholls, K. H. (2013). New and little-known freshwater chrysophytes, including a description of *Randersenia wujeki* gen. et sp. nov. (Chrysophyceae). *Algological Studies, 141*, 53–72.

Pienaar, R. N. (1980). Chrysophytes. In E. R. Cox (Ed.), *Phytoflagellates* (pp. 213–242). Elsevier/North Holland.

Preisig, H. R. (1994). A modern concept of chrysophyte classification. In C. Sandgren, J. Smol, & J. Kristiansen (Eds.), *Chrysophyte algae. Ecology, phylogeny and development* (pp. 46–74). Cambridge University Press.

Sandgren, C. D., Smol, J. P., & Kristiansen, J. (Eds.). (1995). *Chrysophyte algae. Ecology, phylogeny and development* (p. 399). Cambridge University Press.

Siver, P. A., & Lott, A. M. (2017). The scaled chrysophyte flora in freshwater ponds and lakes from Newfoundland, Canada, and their relationship to environmental variables. *Cryptogamie Algologie, 38*(4), 325–347.

Wujek, D., & Carter, S. (2002). Identification, ecology, and distribution of silica-scale bearing chrysophytes from the Carolinas, II. Mountain region. *Journal of the North Carolina Academy of Science, 118*, 137–144.

5 Eustigmatophytes, Xanthophytes, Raphidophytes, and Cryptophytes

The Eustigmatophytes (Class Eustigmatophyceae), Xanthophytes (Class Xanthophyceae), and the Raphidophytes (Class Raphidophyceae), also within the kingdom Chromista, belong to the phylum Heterokontophyta. Cryptophytes are in their own phylum, Cryptista. New genetic information has facilitated the proper taxonomic placement in Eustigmatophytes and Xanthophytes, and many have been reassigned between them. The Eustigmatophytes family consists of only 30 to 33 described genera. They are all unicellular algae and most live in freshwater habitats or damp soils, and no sexual reproduction has been recorded. Eustigmatophytes have chlorophyll *a*, and no other type of chlorophyll, and lack fucoxanthin. They have two dominant xanthophylls, violaxanthin (orange color) and vaucheriaxanthin, which give them a yellow-green appearance, and these pigments are part of the photosynthetic array that captures far-red light that is passed on to chlorophyll *a*. A large red to orange lipid globule is common, as seen in *Vischeria* (*see* Fig. 5.1). A few members of Eustigmatophytes are known to accumulate lipids, making them a valuable candidate for biofuel production. The cell wall is composed mostly of cellulose, but the exact composition has not been determined for most of the organisms in this group.

B. H. Rosen, *Color Atlas of Freshwater Algae*,
https://doi.org/10.1007/978-3-031-93122-2_5

Class: Eustigmatophyceae
Order: Eustigmatales
Family: Chlorobotryaceae

Eustigmatophyceae

Vischeria

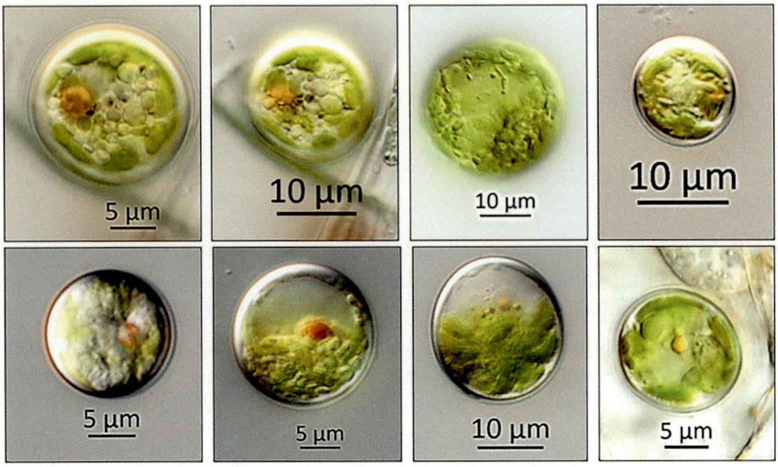

Vischeria polyphem

Metaphyton, epiphyton

Fig. 5.1 *Vischeria*

Class: Eustigmatophyceae
Order: Eustigmatales
Family: Chlorobotryaceae

Eustigmatophyceae

Vischeria

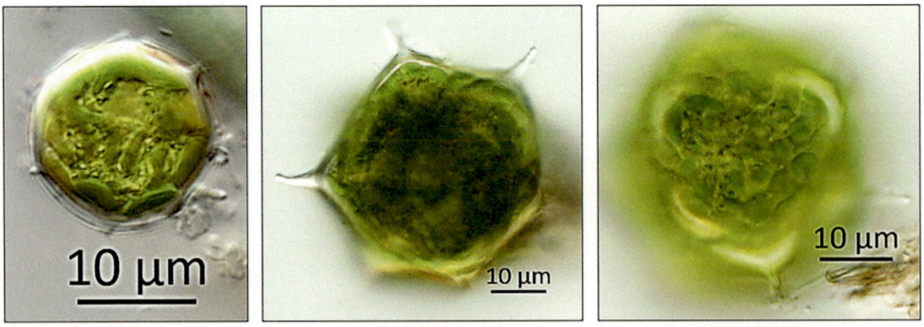

Vischeria stellate

Metaphyton, epiphyton

Fig. 5.2 *Vischeria*

Class: Eustigmatophyceae
Order: Goniochloridales
Family: Goniochloridaceae

Eustigmatophyceae

Tetraëdriella

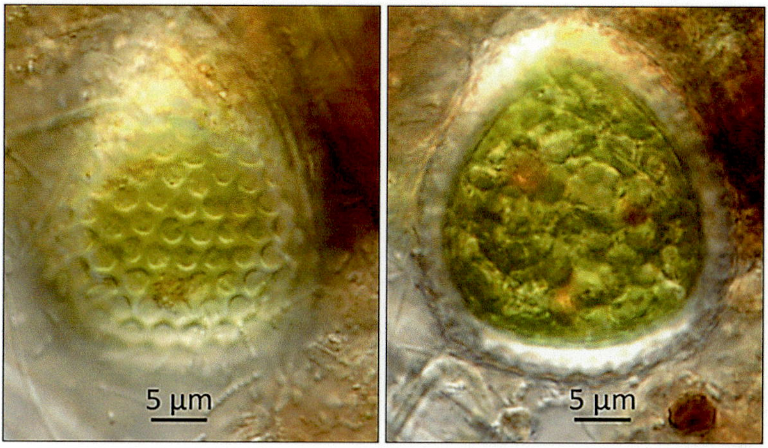

Tetraëdriella polychloris

Metaphyton, epiphyton

Fig. 5.3 *Tetraëdriella*

Class: Eustigmatophyceae
Order: Goniochloridales
Family: Goniochloridaceae

Eustigmatophyceae

Goniochloris

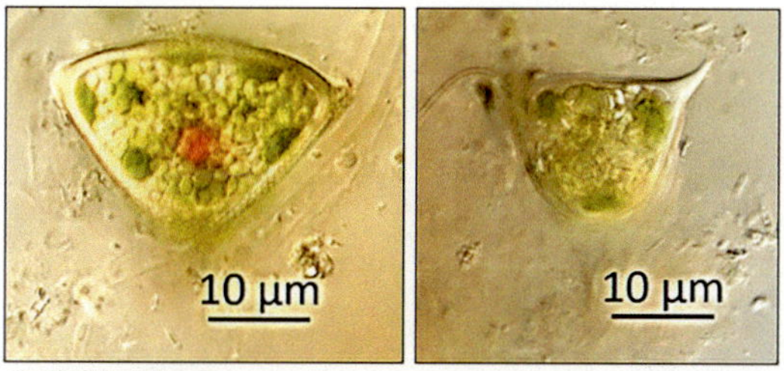

Goniochloris laevis

Goniochloris sculpta

Metaphyton, epiphyton

Fig. 5.4 *Goniochloris*

Class: Eustigmatophyceae
Order: Goniochloridales
Family: Goniochloridaceae

Eustigmatophyceae

Trachydiscus

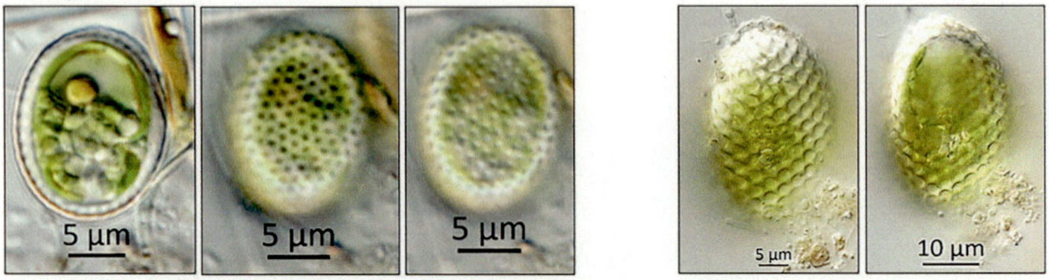

Trachydiscus ellipsoideus *Trachydiscus sexangulatus*

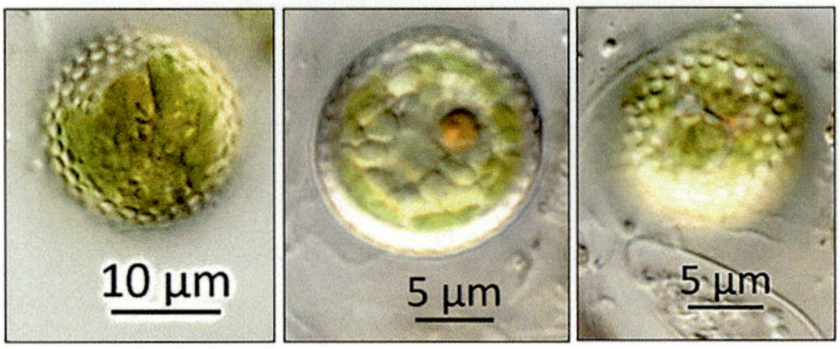

Trachydiscus lenticularis

Metaphyton, Epiphyton

Fig. 5.5 *Trachydiscus*

Class: Eustigmatophyceae
Order: Eustigmatales
Family: Chlorobotryaceae

Eustigmatophyceae

Characiopsis

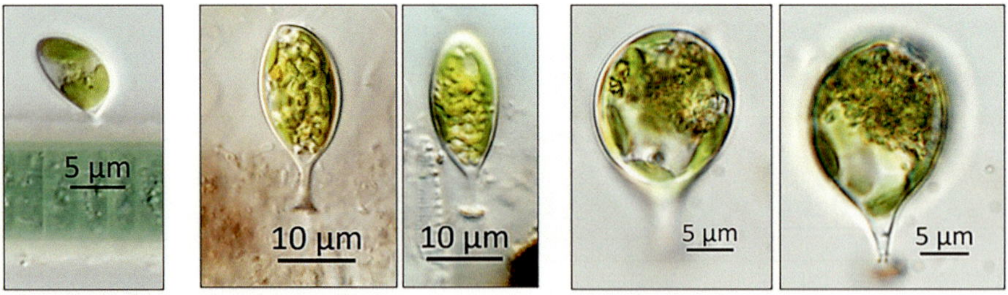

Characiopsis gibba *Characiopsis pyriformis* *Characiopsis malleolus*

Epiphyton

Fig. 5.6 *Characiopsis*

Class: Eustigmatophyceae
Order: Eustigmatales
Family: Chlorobotryaceae

Eustigmatophyceae

Lietzensia

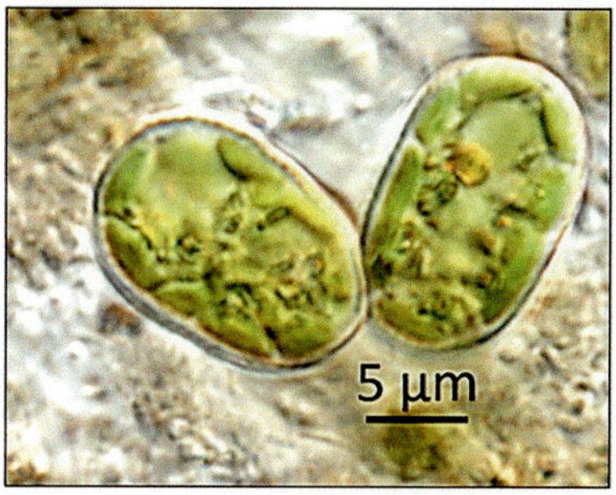

Lietzensia polymorpha

Metaphyton, epiphyton

Fig. 5.7 *Lietzensia*

Class: Eustigmatophyceae
Order: Eustigmatales
Family: Monodopsidaceae

Eustigmatophyceae

Monodopsis

Monodopsis subterranea

Metaphyton, epiphyton

Fig. 5.8 *Monodopsis*

Class: Eustigmatophyceae
Order: Goniochloridales
Family: Gonoiochloridaceae

Eustigmatophyceae

Pseudostaurastrum

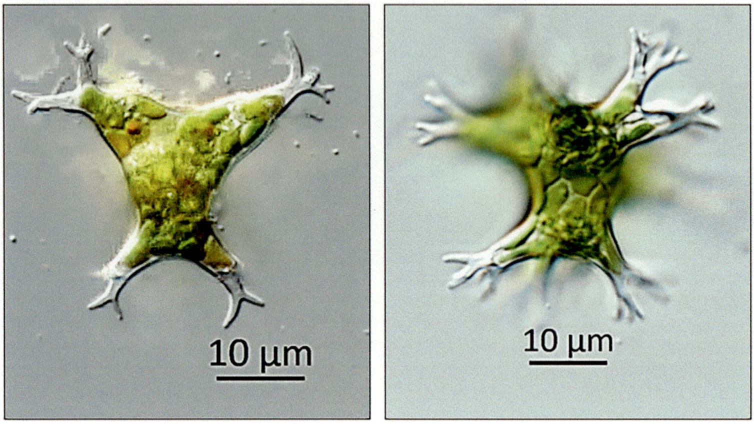

Pseudostaurastrum enorme

Metaphyton, Epiphyton

Fig. 5.9 *Pseudostaurastrum*

The Xanthophytes are also known as yellow-green algae, have chlorophyll *a* and *c* and the photoprotective pigment diadinoxanthin, but like the Eustigmatophytes, they lack fucoxanthin. Most are freshwater, although some are on damp soils or tree trunks or marine, with about 100 genera. Most are nonmotile unicells typically attached (sessile) to other algae or a substrate by mucilaginous materials or a stalk. A few genera are more advanced, forming filaments, and some have pronounced cross-walls between cells, which form a characteristic **H** "piece" from the ends of two adjacent cells (*Tribonema*, see Fig. 5.20). Another distinguishing feature of the Xanthophytes is yellow-green discoid chloroplast. One genus, *Vaucheria*, lacks cross-walls altogether, except when branching (see Fig. 5.22) or making reproductive structures. The lack of cross-walls is termed "coenocytic," and in this genus, each cell contains multiple nuclei and discoid chloroplasts.

Class: Xanthophyceae
Order: Mischococcales
Family: Pleurochloridaceae

Xanthophytes

Isthmochloron

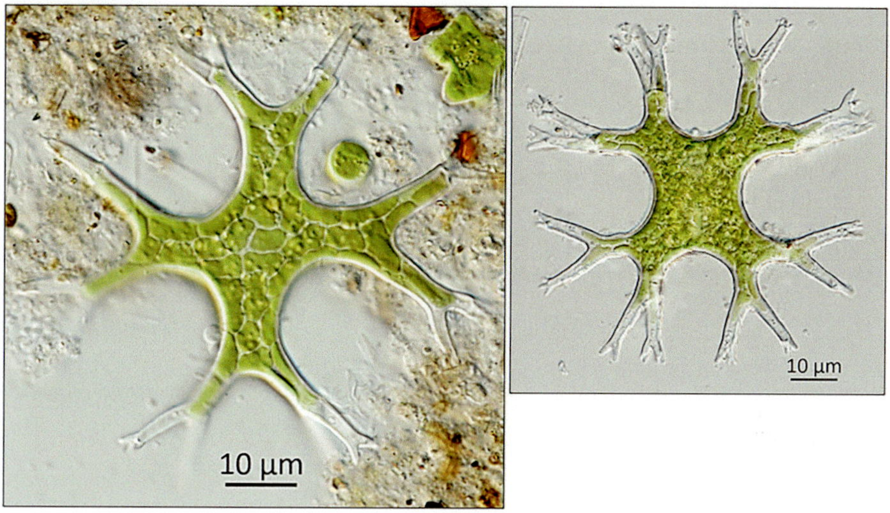

Isthmochloron lobulatum

Metaphyton

Fig. 5.10 *Isthmochloron*

Class: Xanthophyceae
Order: Mischococcales
Family: Botrydiopsisdaceae

Xanthophytes

Botrydiopsis

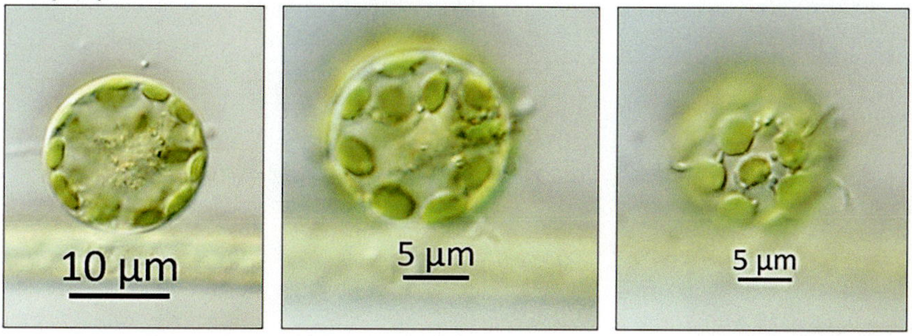

Botrydiopsis arhiza

Metaphyton, Epiphyton

Fig. 5.11 *Botrydiopsis*

Class: Xanthophyceae
Order: Mischococcales
Family: Pleurochloridaceae

Xanthophyceae

Monodus

Monodus chodatii

Monallantus

Monallantus pyreniger

Metaphyton, epiphyton

Fig. 5.12 *Monodus* and *Monallantus*

Class: Xanthophyceae
Order: Mischococcales
Family: Pleurochloridaceae

Xanthophyceae

Tetraplektron

Tetraplektron acutum

Pseudogoniochloris

Pseudogoniochloris tripus

Metaphyton, Epiphyton

Fig. 5.13 *Tetraplektron* and *Pseudogoniochloris*

Class: Xanthophyceae
Order: Mischococcales
Family: Sciadiaceae

Xanthophytes

Centritractus

Centritractus belonophorus Plankton, metaphyton

Fig. 5.14 *Centritractus*

Class: Xanthophyceae
Order: Mischococcales
Family: Sciadiaceae

Xanthophyceae

Bumilleriopsis

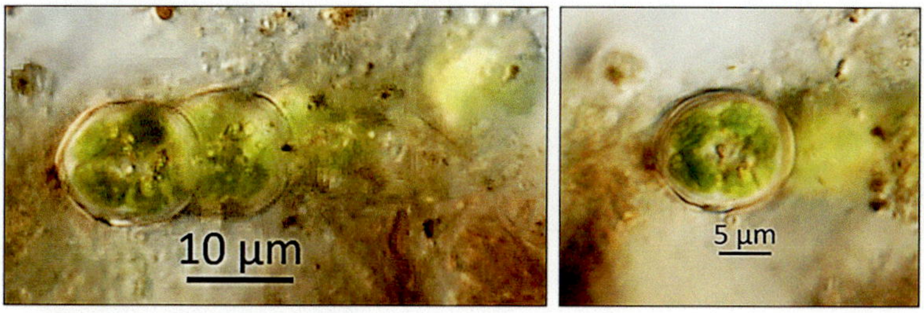

Bumilleriopsis filiformis Metaphyton, Epiphyton

Fig. 5.15 *Bumilleriopsis*

Class: Xanthophyceae
Order: Mischococcales
Family: Sciadiaceae

Xanthophyceae

Chlorothecium

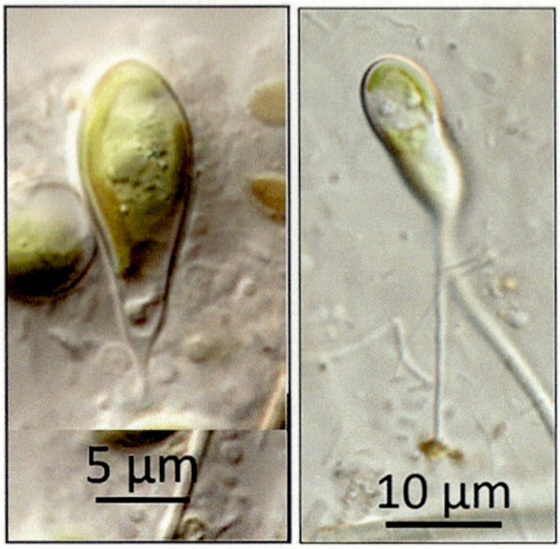

Chlorothecium clava

Epiphyton

Fig. 5.16 *Chlorothecium*

Class: Xanthophyceae
Order: Mischococcales
Family: Sciadiaceae

Xanthophyceae

Ophiocytium

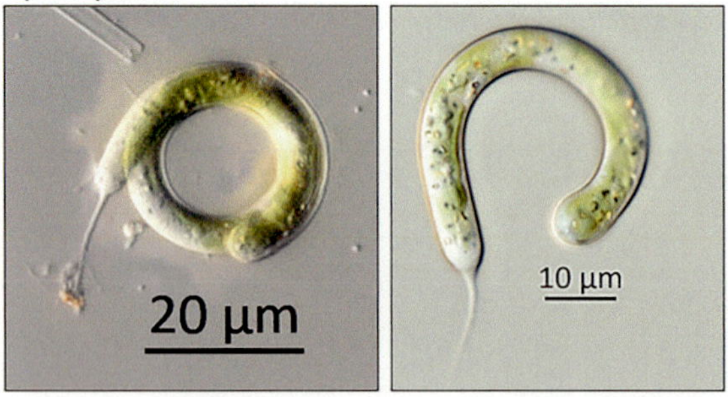

Ophiocytium cochleare

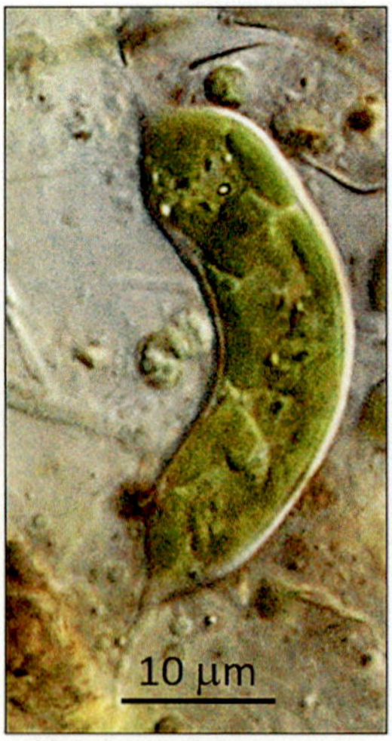

Ophiocytium capitatum

Metaphyton, epiphyton

Fig. 5.17 *Ophiocytium*

Class: Xanthophyceae
Order: Mischococcales
Family: Sciadiaceae

Xanthophyceae

Ophiocytium

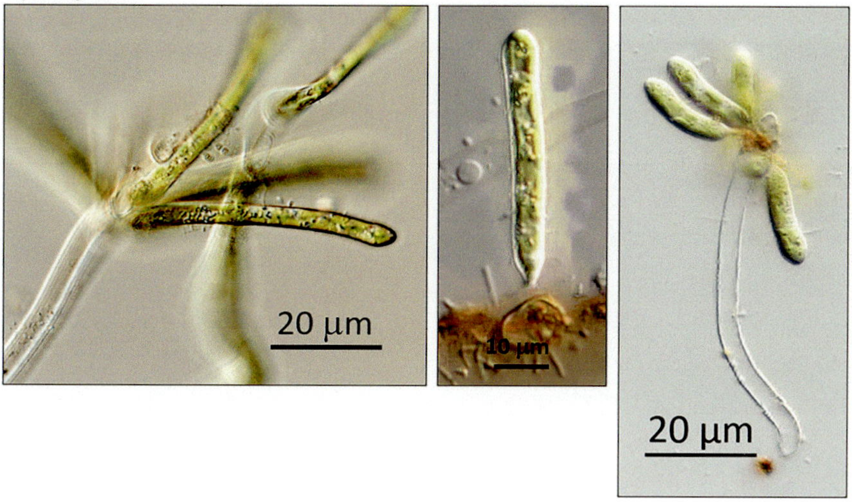

Ophiocytium arbuscula

Metaphyton, epiphyton

Fig. 5.18 *Ophiocytium*

Class: Xanthophyceae
Order: Mischococcales
Family: Sciadiaceae

Xanthophyceae

Ophiocytium

Ophiocytium parvulum

Metaphyton, Epiphyton

Fig. 5.19 *Ophiocytium*

Class: Xanthophyceae
Order: Tribonematales
Family: Tribonemataceae

Xanthophytes

Tribonema

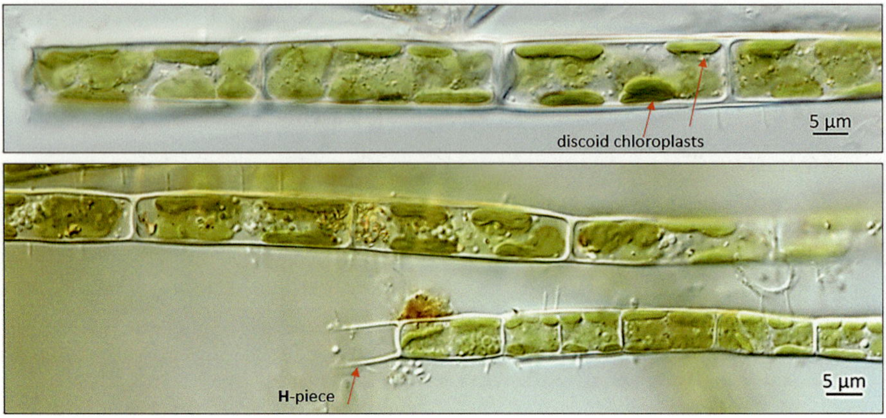

Tribonema vulgare Metaphyton

Fig. 5.20 *Tribonema*

Class: Xanthophyceae
Order: Vaucheriales
Family: Vaucheriaceae

Xanthophyceae

Vaucheria

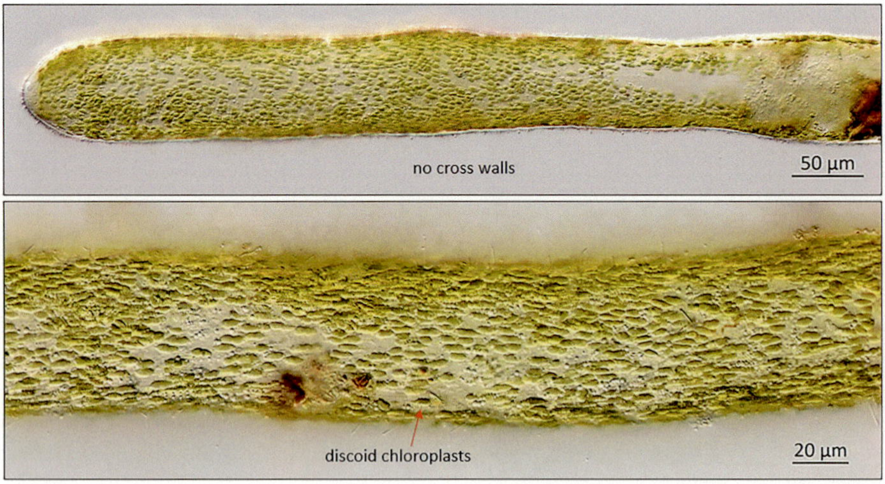

Vaucheria tuberosa

Fig. 5.21 *Vaucheria*

Class: Xanthophyceae
Order: Vauchariales
Family: Vaucheriaceae

Xanthophyceae

Vaucheria

Vaucheria tuberosa

Metaphyton

Fig. 5.22 *Vaucheria*

Class: Xanthophyceae
Order: Xanthophyceae *ordo incertae sedis*
Family: Xanthophyceae *familia incertae sedis*

Xanthophyceae

Dioxys

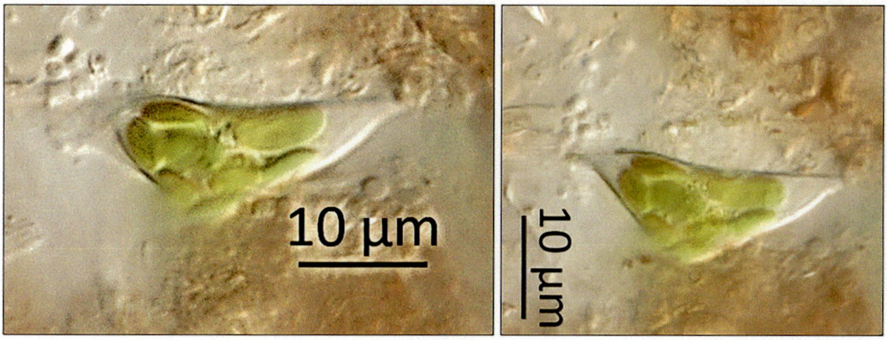

Dioxys incus

Metaphyton, epiphyton

Fig. 5.23 *Dioxys*

Class: Xanthophyceae
Order: Xanthophyceae *ordo incertae sedis*
Family: Xanthophyceae *familia incertae sedis*

Xanthophyceae

Dioxys

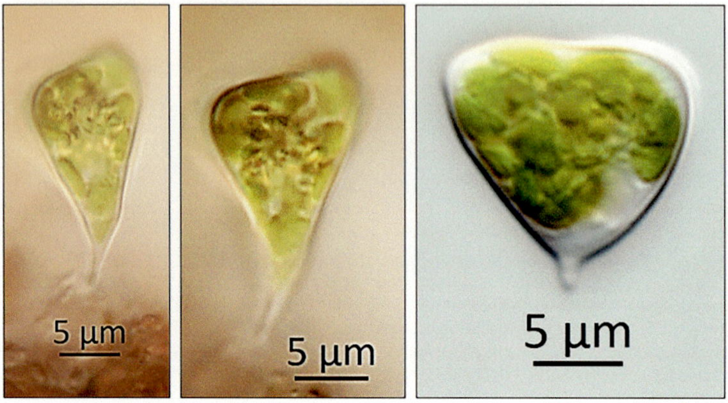

Dioxys inermis

Metaphyton, epiphyton

Fig. 5.24 *Dioxys*

Class: Xanthophyceae
Order: Xanthophyceae *ordo incertae sedis*
Family: Xanthophyceae *familia incertae sedis*

Xanthophyceae

Lutherella

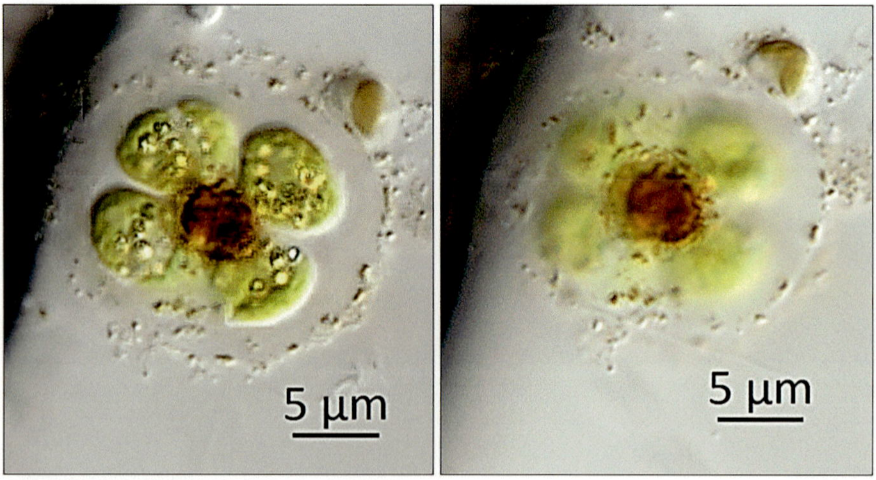

Note: the Order and Family of this genus are uncertain

Lutherella obovoidea

Metaphyton, epiphyton

Fig. 5.25 *Lutherella*

The Raphidophytes are a small class of flagellated unicells and no cell wall; they are simply surrounded by the cell membrane. One flagellum projects forward and the other trailing. Without the cell wall, these organisms can readily change their shape (metaboly) as they are swimming. The chloroplasts contain chlorophyll *a* and *c* and vaucheriaxanthin and diadinoxanthin in freshwater genera.

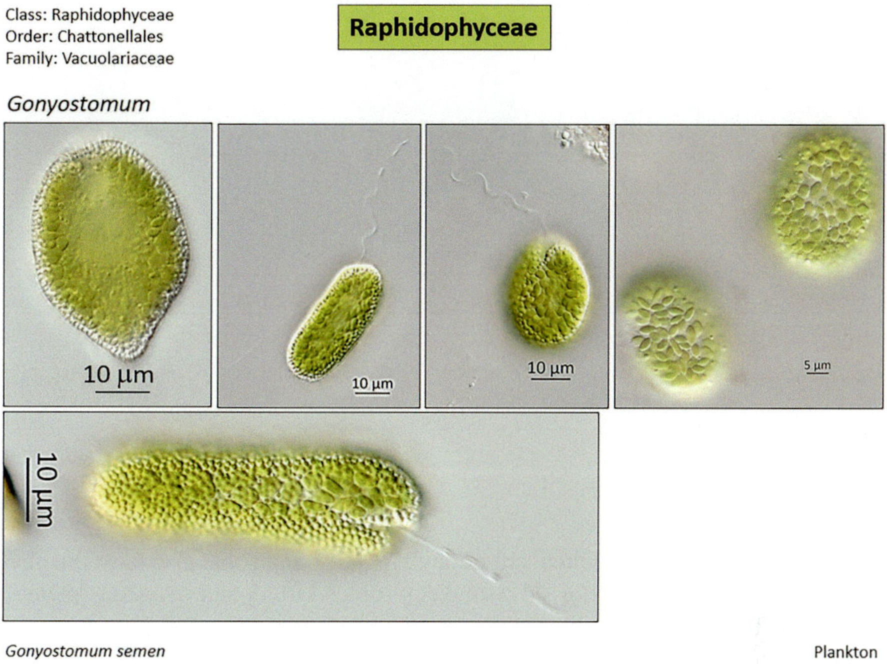

Fig. 5.26 *Gonyostomum*

Class: Raphidophyceae
Order: Chattonellales
Family: Vacuolariaceae

Raphidophyceae

Gonyostomum

Gonyostomum latum

Plankton

Fig. 5.27 *Gonyostomum*

Cryptophyte pigments include chlorophyll *a* and *c*, and some also have the phycobilins. They have a gullet, lined with ejectosomes killing and digesting bacteria and protists (Figs. 5.28 to 5.29).

Class: Cryptophyceae
Order: Cryptomonadales
Family: Cryptomonadaceae

Cryptophytes

Cryptomonas

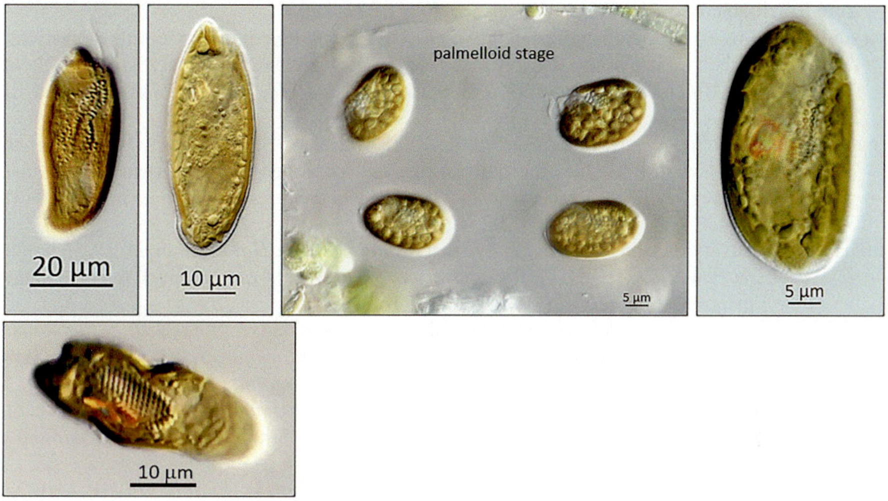

Cryptomonas ovata

Plankton

Fig. 5.28 *Cryptomonas*

Class: Cryptophyceae
Order: Pyrenomonadales
Family: Hemiselmidaceae

Cryptophytes

Chroomonas

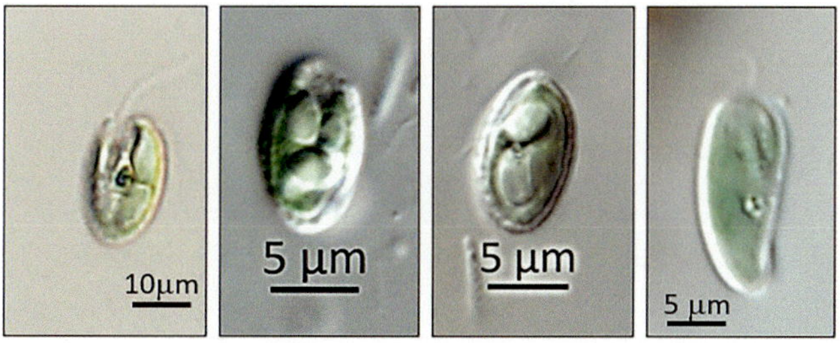

Chroomonas nordstedtii

Plankton

Fig. 5.29 *Chroomonas*

Suggested Reading

Branco, S., Almeida, L. L., Alves-de-Souza, C., Oliveira, M. M. M., Proença, L. A. O., & Menezes, M. (2019). Morphological and genetic characterization of bloom-forming *Raphidophyceae* from Brazilian coast. *Phycological Research, 67*(4), 279–290.

Diesing, K. M. (1866). Revision der Prothelminthen. Abtheilung: Mastigophoren. Sitzungsberichte der Kaiserlichen Akademie der Wissenschaften. Mathematisch-Naturwissenschaftliche Classe. Abt. 1. *Mineralogie, Botanik, Zoologie, Anatomie, Geologie und Paläontologie, 52*(8), 287–401.

Engesmo, A., Eikrem, W., Seoane, S., Smith, K., Edvardsen, B., Hofgaard, A., & Tomas, C. R. (2016). New insights into the morphology and phylogeny of *Heterosigma akashiwo* (Raphidophyceae), with the description of *Heterosigma minor* sp. nov. *Phycologia, 55*(3), 279–294.

Ettl, H. (1977). Taxonomische Bemerkungen zu den Xanthophyceen. *Nova Hedwigia, 28*, 555–568.

Fawley, M. W., & Fawley, K. P. (2017). Rediscovery of *Tetraëdriella subglobosa* Pascher, a member of the Eustigmatophyceae. *Fottea, Olomouc, 17*(1), 96–102.

Fritsch, F. E. (1935). The structure and reproduction of the algae. *Volume I. Introduction, Chlorophyceae. Xanthophyceae, Chrysophyceae, Bacillariophyceae, Cryptophyceae, Dinophyceae, Chloromonineae, Euglenineae, Colourless Flagellata*. (pp. i–xvii, 1–791).

Hegewald, E., Padisak, J., & Friedl, T. (2007). *Pseudotetraedriella kamillae*: Taxonomy and ecology of a new member of the algal class Eustigmatophyceae (Stramenopiles). *Hydrobiologia, 586*, 107–116.

Hill, D. R. A. (1991). *Chroomonas* and other blue-green cryptomonads. *Journal of Phycology, 27*, 133–145.

Neustupa, J. (2002). An easy method for light microscopic distinguishing of xanthophycean and eustigmatophycean strains. *Fottea, 2*(1), 43–46.

Pascher, A. (1939). Heterokonten. In L. Rabenhorst (Ed.), *Kryptogamen-Flora von Deutschland, Österreich und der Schweiz* (Vol. 11 Teil 6, pp. 833–1092).

Rieth, A. (1980). Xanthophyceae. In: Süsswasserflora von Mitteleuropa (Eds.) Vol.4(1), pp. 1–147. Gustav Fischer.

Schiller, J. (1926). Der thermische Einfluß und die Wirkung des Eises auf die planktischen Herbstvegetationen in den Altwässern der Donau bei Wien nach regelmäßiger Beobachtung von Oktober 1918 bis Ende 1925. *Archiv für Protistenkunde, 56*(1), 1–62.

Smith, T. E. (2010). Revised list of algae from Arkansas, U.S.A. and new additions. *International Journal on Algae, 12*(3), 230–256.

Euglenophytes

6

The phylum Euglenozoa is placed in the kingdom Protozoa (Cavalier-Smith, 2016), and only the class Euglenophyceae in the phylum Euglenozoa acquired green algal chloroplasts—a process called symbiogenesis. Because of the symbiotic origin, the chloroplast is surrounded by an extra membrane. They have both chlorophyll *a* and *b,* making the organisms grass-green in color. They also have betacarotene, diadinoxanthin, and as in *Euglena sanguinea* (*see* Fig. 6.4), the red pigment astaxanthin. All euglenoids can use sunlight as an energy source, but some are also facultative heterotrophs, obtaining energy and nutrients by consuming other organisms or organic compounds in the water. The morphology of the chloroplasts varies widely, from discoid to ribbons, and the number per cell can range from a couple to several hundreds. All these features are used to identify the organism. Most of the euglenoids are highly motile, with two flagella that originate in a subapical groove. Most genera have one long flagellum that emerges from the groove and provides helical movement of the cells. The other flagellum is typically confined to the groove. All also have a red-orange eyespot (b-carotene) and a strong phototactic response to light.

Within the cell membrane is the pellicle: a matrix of flexible, interlocking protein strips that twist throughout the length of the cell and give the organism a striped appearance. In some species, the pellicle is ridged and, as an organism swims, it appears to simply twist through the water, such as those in the genus *Phacus* (*see* Fig. 6.14). Other genera are more flexible, and the entire cell alters its shape as it swims or moves through the microhabitats.

The carbohydrates produced during photosynthesis are stored as paramylon, b-1,3-linked glucose molecules and act as a food reserve and may play a role in chemical storage and ion exchange. In many genera, the shape and quantity of paramylon "bodies" are easily discernible under the microscope and are an identification feature.

B. H. Rosen, *Color Atlas of Freshwater Algae*,
https://doi.org/10.1007/978-3-031-93122-2_6

One common genus, *Trachelomonas*, produces a spherical to cylindrical covering, made of minerals and mucilage, called the lorica. The lorica completely encases the cell, except for a small apical opening, often collared, that provides access for the long flagellum. The lorica may be smooth, rough, and have other ornamentation, such as spines. The patterns of the lorica are used for taxonomic identification of species. The genus *Strombomonas* also produces a lorica; however, it is tapered at the end (Figs. 6.17 and 6.18).

Class: Euglenophyceae
Order: Euglenales
Family: Euglenaceae

Euglenoids

Colacium

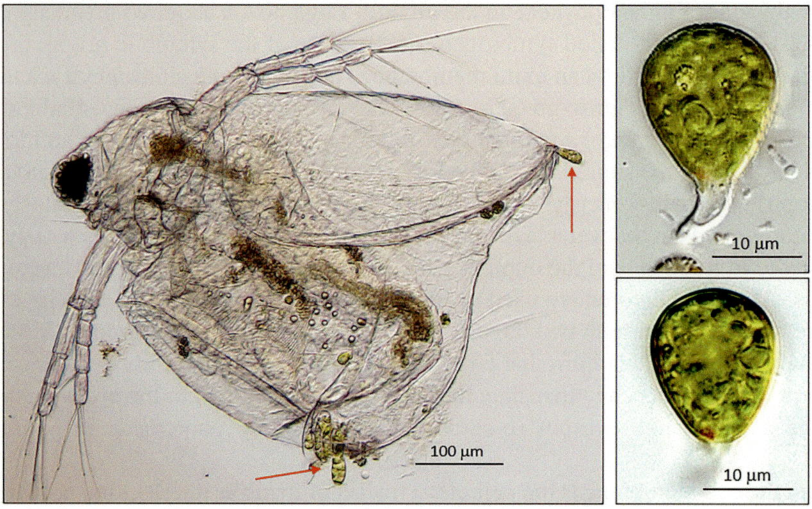

Colacium vesiculosum

Epizoic on *Daphnia* and other invertebrates or filamentous algae

Fig. 6.1 *Colacium*

Class: Euglenophyceae
Order: Euglenales
Family: Euglenaceae

Euglenoids

Cryptoglena

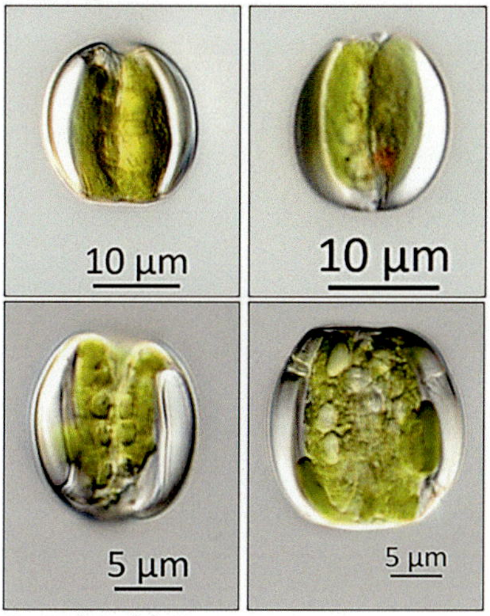

Cryptoglena skujae

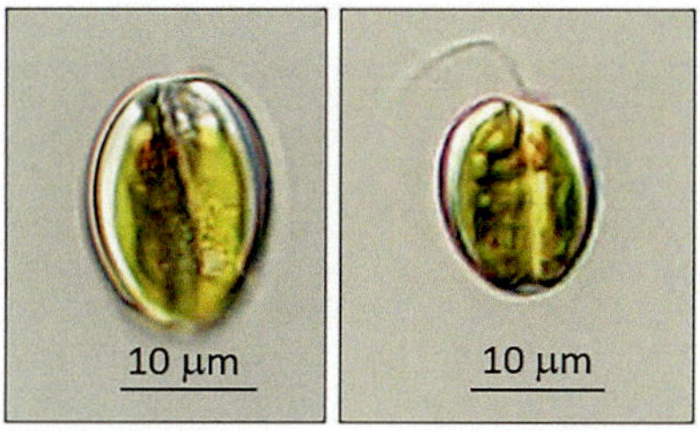

Cryptoglena pigra

Plankton, metaphyton

Fig. 6.2 *Cryptoglena*

Class: Euglenophyceae
Order: Euglenales
Family: Euglenaceae

Euglenoids

Euglena

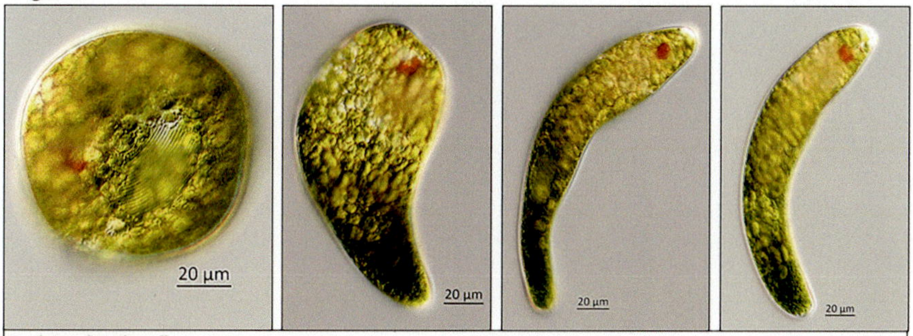

Euglena ehrenbergii Sequential images

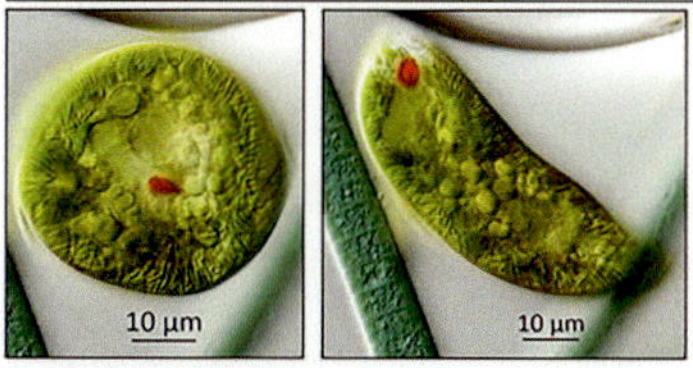

Euglena splendens

Plankton, metaphyton

Fig. 6.3 *Euglena*

Class: Euglenophyceae
Order: Euglenales
Family: Euglenaceae

Euglenoids

Euglena

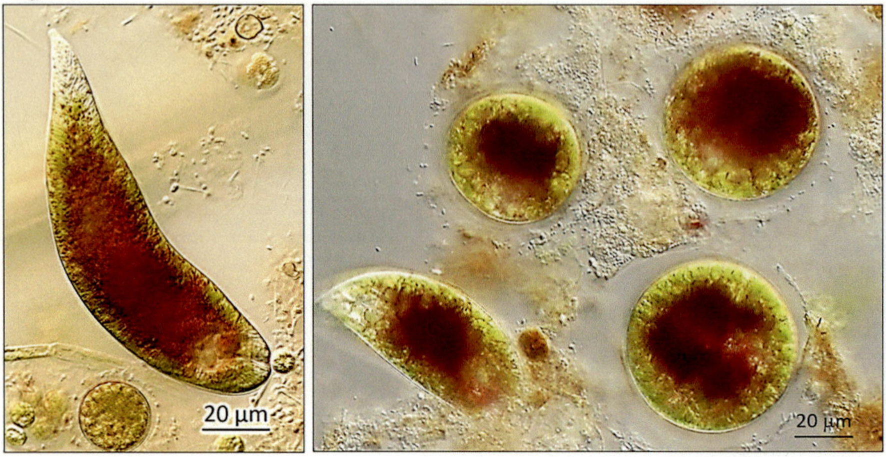

Euglena sanguinea

Plankton, metaphyton

Fig. 6.4 *Euglena*

Class: Euglenophyceae
Order: Euglenales
Family: Euglenaceae

Euglenoids

Euglenaformis

Euglenaformis proxima

Plankton, metaphyton

Fig. 6.5 *Euglenaformis*

Class: Euglenophyceae
Order: Euglenales
Family: Euglenaceae

Euglenoids

Flexiglena

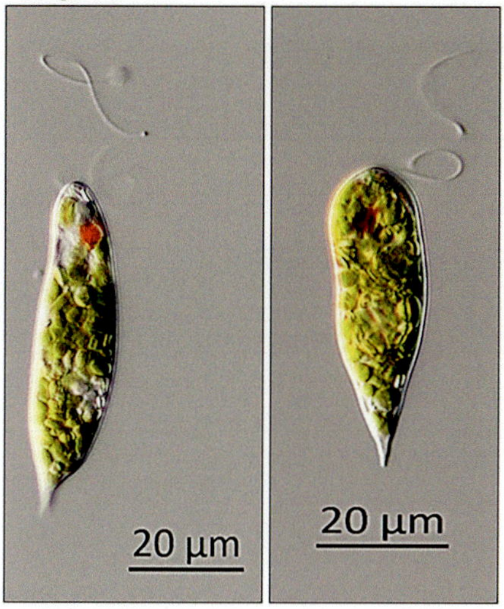

Flexiglena variabilis

Plankton, metaphyton

Fig. 6.6 *Flexiglena*

Class: Euglenophyceae
Order: Euglenales
Family: Euglenaceae

Euglenoids

Flexiglena

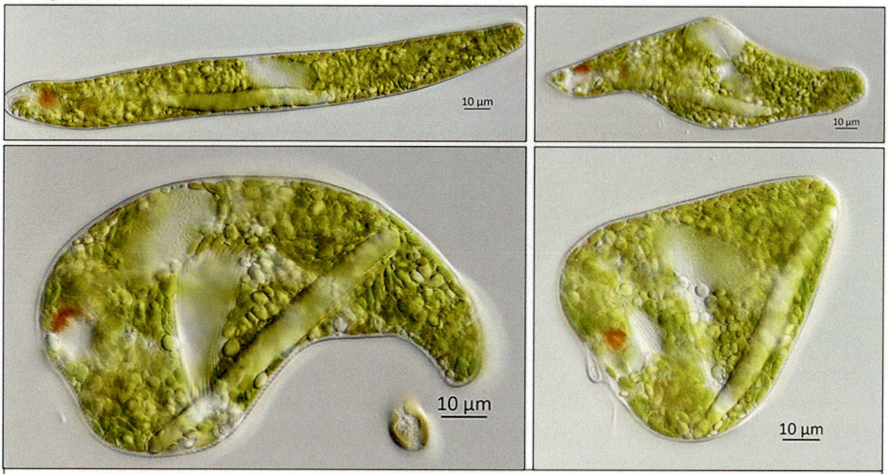

Sequential images

Flexiglena variabilis

Plankton, metaphyton

Fig. 6.7 *Flexiglena*

Class: Euglenophyceae
Order: Euglenales
Family: Euglenaceae

Euglenoids

Lepocinclis

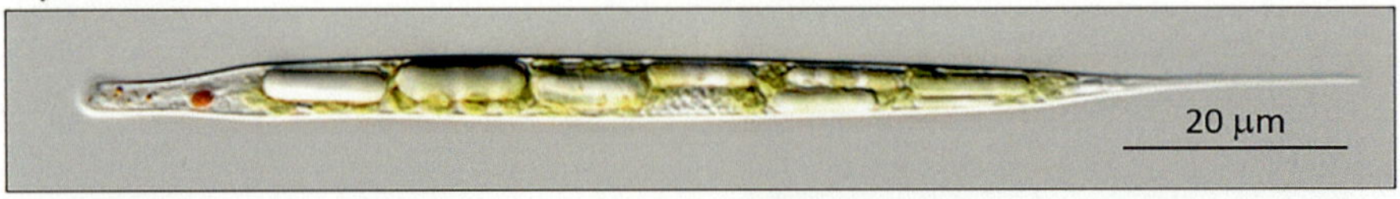

Lepocinclis acus

Lepocinclis longissimi

Periphyton, metaphyton

Fig. 6.8 *Lepocinclis*

Class: Euglenophyceae
Order: Euglenales
Family: Euglenaceae

Euglenoids

Lepocinclis

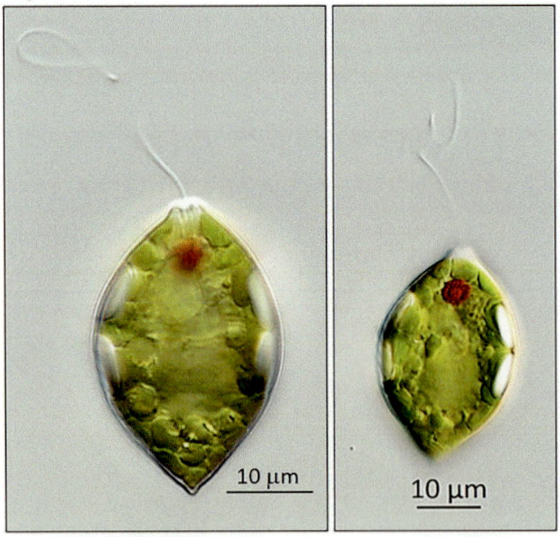

Lepocinclis fusiformis

Lepocinclis hispidula

Plankton, metaphyton

Fig. 6.9 *Lepocinclis*

Class: Euglenophyceae
Order: Euglenales
Family: Euglenaceae

Euglenoids

Lepocinclis

Lepocinclis steinii

Plankton, metaphyton

Fig. 6.10 *Lepocinclis*

Class: Euglenophyceae
Order: Euglenales
Family: Euglenaceae

Euglenoids

Lepocinclis

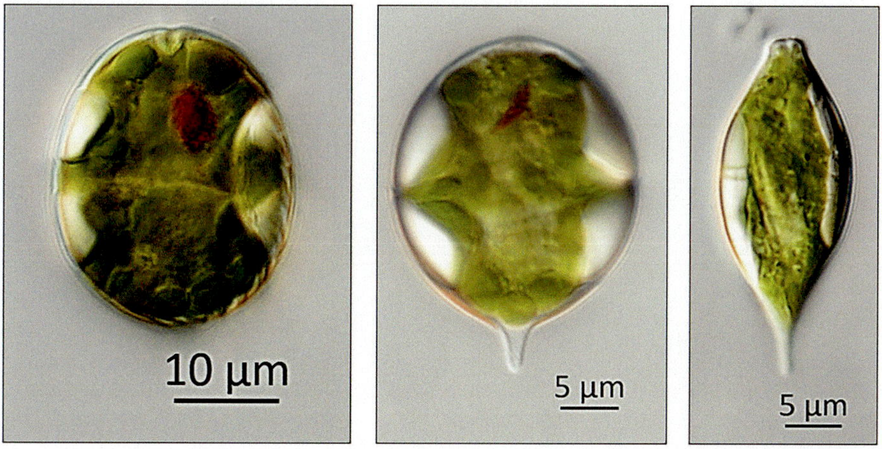

Lepocinclis globulus *Lepocinclis ovum* *Lepocinclis autumnalis*

Plankton, metaphyton

Fig. 6.11 *Lepocinclis*

Class: Euglenophyceae
Order: Euglenales
Family: Euglenaceae

Euglenoids

Lepocinclis

Lepocinclis fusca

Lepocinclis sp.

Plankton, metaphyton

Fig. 6.12 *Lepocinclis*

Class: Euglenophyceae
Order: Euglenales
Family: Euglenaceae

Euglenoids

Monomorphina

Monomorphina cf. *pyrum*

Plankton, metaphyton

Fig. 6.13 *Monomorphina*

Class: Euglenophyceae
Order: Euglenales
Family: Euglenaceae

Euglenoids

Phacus

Phacus cristatus

Plankton, metaphyton

Fig. 6.14 *Phacus*

Class: Euglenophyceae
Order: Euglenales
Family: Euglenaceae

Euglenoids

Phacus

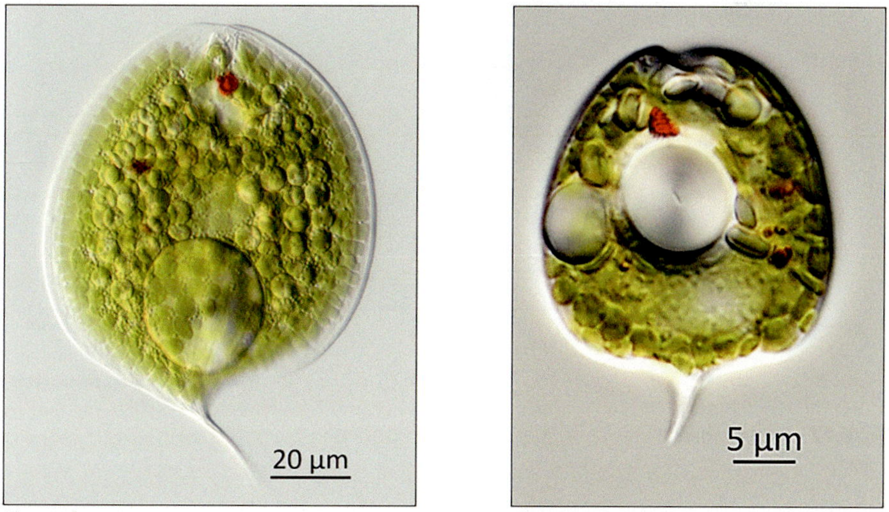

Phacus gigas *Phacus pleuronectes*

Fig. 6.15 *Phacus*

Class: Euglenophyceae
Order: Euglenales
Family: Euglenaceae

Euglenoids

Phacus

Phacus polytrophos

Plankton, metaphyton

Fig. 6.16 *Phacus*

Class: Euglenophyceae
Order: Euglenales
Family: Euglenaceae

Euglenoids

Strombomonas

Strombomonas acuminata

Strombomonas triquetra

Plankton, metaphyton

Fig. 6.17 *Strombomonas*

Class: Euglenophyceae
Order: Euglenales
Family: Euglenaceae

Euglenoids

Strombomonas

Strombomonas maxima

Plankton, metaphyton

Fig. 6.18 *Strombomonas*

Class: Euglenophyceae
Order: Euglenales
Family: Euglenaceae

Euglenoids

Trachelomonas

Trachelomonas armata

Plankton, metaphyton

Fig. 6.19 *Trachelomonas*

Class: Euglenophyceae
Order: Euglenales
Family: Euglenaceae

Euglenoids

Trachelomonas

Trachelomonas spp. Plankton, metaphyton

Fig. 6.20 *Trachelomonas*

Class: Euglenophyceae
Order: Euglenales
Family: Euglenaceae

Euglenoids

Trachelomonas

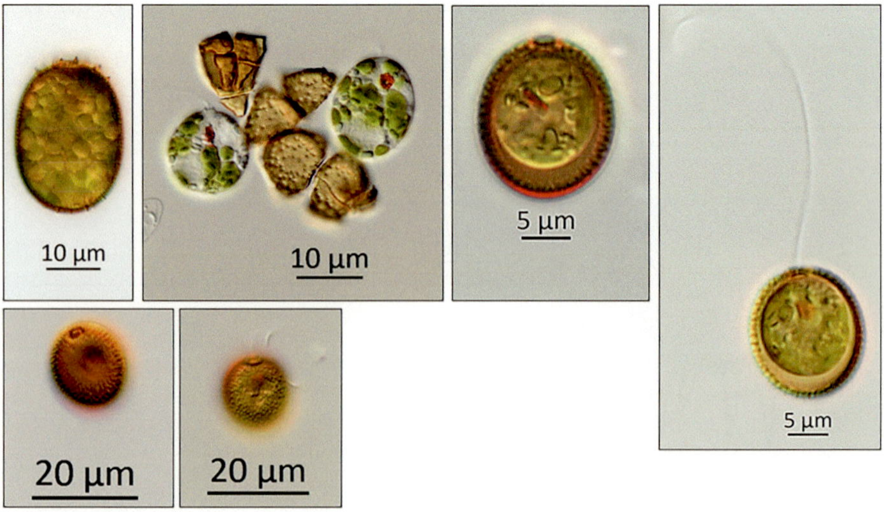

Trachelomonas spp.

Fig. 6.21 *Trachelomonas*

Class: Euglenophyceae
Order: Euglenales
Family: Euglenaceae

Euglenoids

Trachelomonas

Trachelomonas klebsii

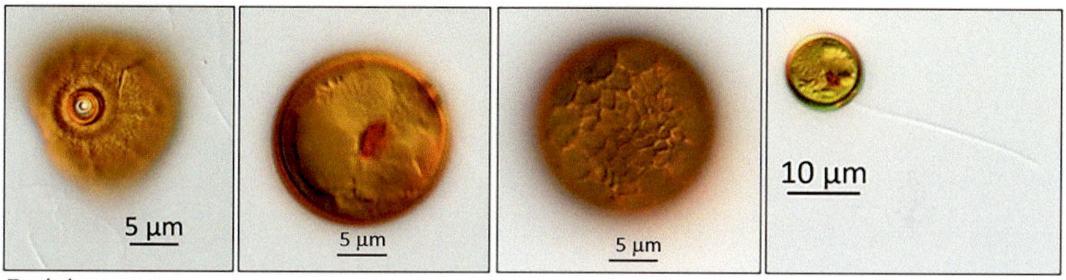

Trachelomonas spp.

Plankton, metaphyton

Fig. 6.22 *Trachelomonas*

Suggested Reading

Chaber, K., Lukomska-Kowalczyk, M., Fells, A., Milanowski, R., & Zakrys, B. (2022). Toward the robust resolution of taxonomic ambiguity within *Lepocinclis* (Euglenida) based on DNA sequencing and morphology. *Journal of Phycology, 58*(1), 105–120.

Jankowska, K., Łukomska-Kowalczyk, M., Milanowski, R., Fells, A., & Zakrys, B. (2024). Biodiversity of autotrophic euglenids based on the group specific DNA metabarcoding approach. *Protist, 175*(126024), 1–13.

Karnkowska-Ishikawa, A., Milanowski, R., Triemer, R. E., & Zakrys, B. (2012). Taxonomic revisions of morphologically similar species from two euglenoid genera: *Euglena* (*E. granulata* and *E. velata*) and *Euglenaria* (*Eu. anabaena, Eu. caudata*, and *Eu. clavata*). *Journal of Phycology, 48*(3), 729–739.

Kosmala, S., Karnkowska-Ishikawa, A., Milanowski, R., Kwiatowski, J., & Zakrys, B. (2009). Phylogeny and systematics of *Euglena* (Euglenaceae) species with axial, stellate chloroplasts based on morphological and molecular data – New taxa, emended diagnoses, and epitypifications. *Journal of Phycology, 45*(2), 464–481.

Lukomska-Kowalczyk, M., Karnkowsak, A., Krupska, M., Milanowski, R., & Zakrys, B. (2016). DNA barcoding in autotrophic euglenids: Evaluation of COI and 18S rDNA. *Journal of Phycology, 52*(6), 951–960.

Lukomska-Kowalczyk, M., Chaber, K., Fells, A., Milanowski, R., & Zakrys, B. (2021). Description of Flexiglena gen. nov. and new members of Discoplastis and Euglenaformis (Euglenida). *Journal of Phycology, 57*(3), 766–779.

Milanowski, R., Kosmala, S., Zakrys, B., & Kwiatowski, J. (2006). Phylogeny of photosynthetic euglenophytes based on combined chloroplast and cytoplasmic SSU rDNA sequence analysis. *Journal of Phycology, 42*, 721–730.

Zakrys, B., Moraczewski, I., & Kucharski, R. (1997). The species concept in Euglena in the light of DNA polymorphism analyses. *Algological Studies/Archiv für Hydrobiologie, Supplement Volumes, 86*, 51–79.

Dinoflagellates 7

The Dinoflagellates (phylum Dinoflagellata) are also within the kingdom Chromista. Dinoflagellates are mostly marine organisms with relatively few freshwater genera. They are mostly planktonic, single-celled organisms; also some live attached to other algae. Some marine species form harm algal blooms in marine systems, such as *Karenia brevis* (*see* Fig. 7.7), associated with Red Tide. Some are also endosymbionts in corals.

Dinoflagellates have two distinct flagella, one that moves the organism forward with a spinning or twirling motion, while the other works as a rudder. The trailing longitudinal flagellum fits into a groove, and the sulcus and the transverse flagellum coil in the cingulum.

Most dinoflagellates are photosynthetic, and because this group readily acquired chloroplasts from other algae through secondary and tertiary symbiosis, it has chlorophyll *a* and *c*. There are some with chlorophyll *b* and phycobilins (with phycocyanin and/or phycoerythrin). Some also have fucoxanthin, dinoxanthin, and most have peridinin as the major carotenoid in their chloroplast.

Some dinoflagellates are purely photosynthetic, while others are mixotrophic, using a variety of energy and carbon sources. Photosynthate is stored as starch, starch-like compounds, or oils. Photosynthesis is achieved either from their own chloroplasts or those that have become endosymbionts acquired through kleptoplasty. This can be seen in *Nusuttodinium aeruginosum* (*see* Fig. 7.4), which acquired phycobilisomes from a Cryptophyte. Some dinoflagellates are predatory on other organisms, including algae.

Most dinoflagellates have a cell wall composed of cellulosic plates embedded within vessels underneath the cell membrane, forming an armor. These are called "armored dinoflagellates." The arrangement of these plates is species specific, and the junction between these plates is a suture. The upper portion of a cell is called the epitheca and may have an apical horn, and the lower portion, the hypotheca, may have antapical horns. Other dinoflagellates lack armor and are called "naked dinoflagellates."

Dinoflagellates have a large nucleus with readily visible chromosomes because they are permanently condensed throughout their life cycle, a unique feature among eukaryotic organisms. There are also a high number of genes and redundant copies of these genes and molecular unknowns with this group of algae (Figs. 7.1 to 7.20).

B. H. Rosen, *Color Atlas of Freshwater Algae*,
https://doi.org/10.1007/978-3-031-93122-2_7

Class: Dinophyceae
Order: Gonyaulacales
Family: Tovelliaceae

Dinoflagellates

Bernardinium

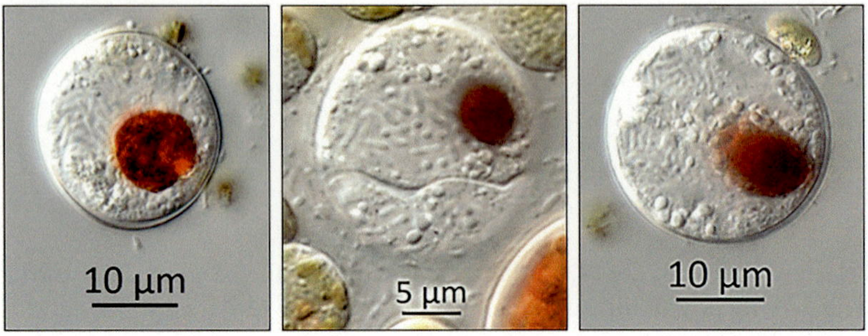

Bernardinium bernardinense Plankton, metaphyton

Fig. 7.1 *Bernardinium*

Class: Dinophyceae
Order: Gonyaulacales
Family: Ceratiaceae

Dinoflagellates

Ceratium

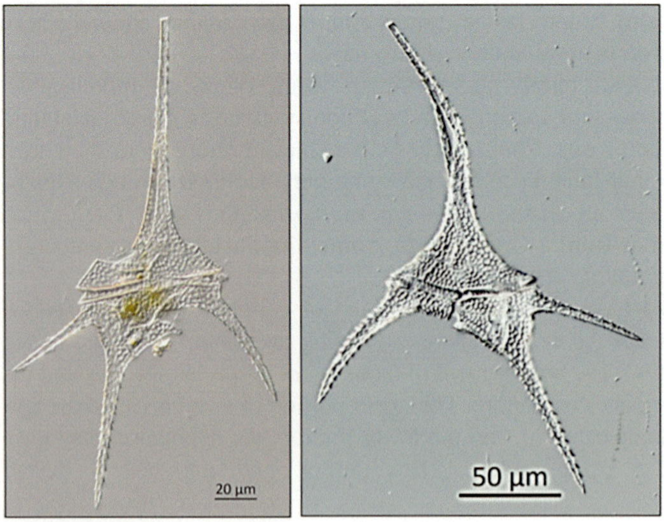

Ceratium hirundinella Plankton

Fig. 7.2 *Ceratium*

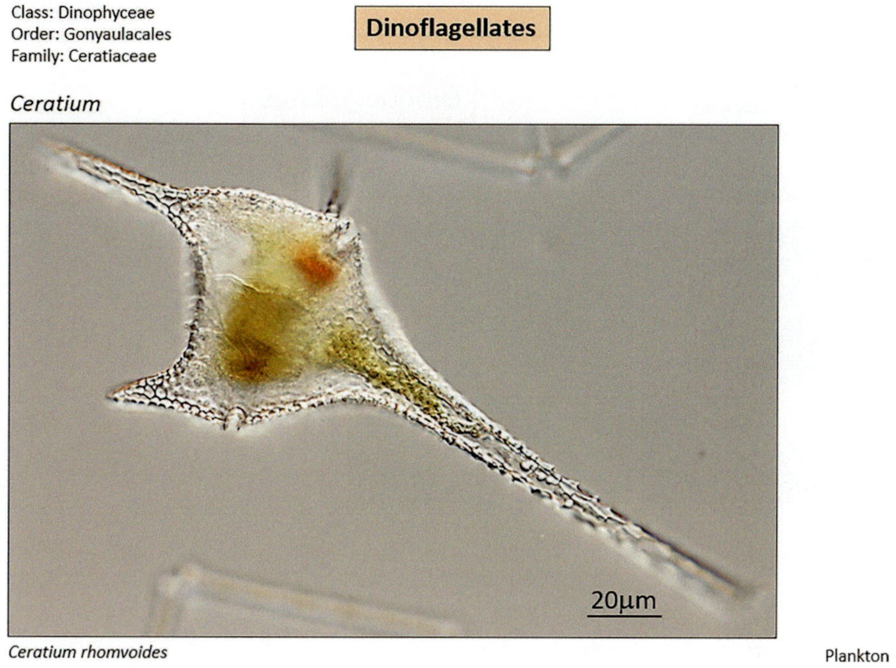

Fig. 7.3 *Ceratium*

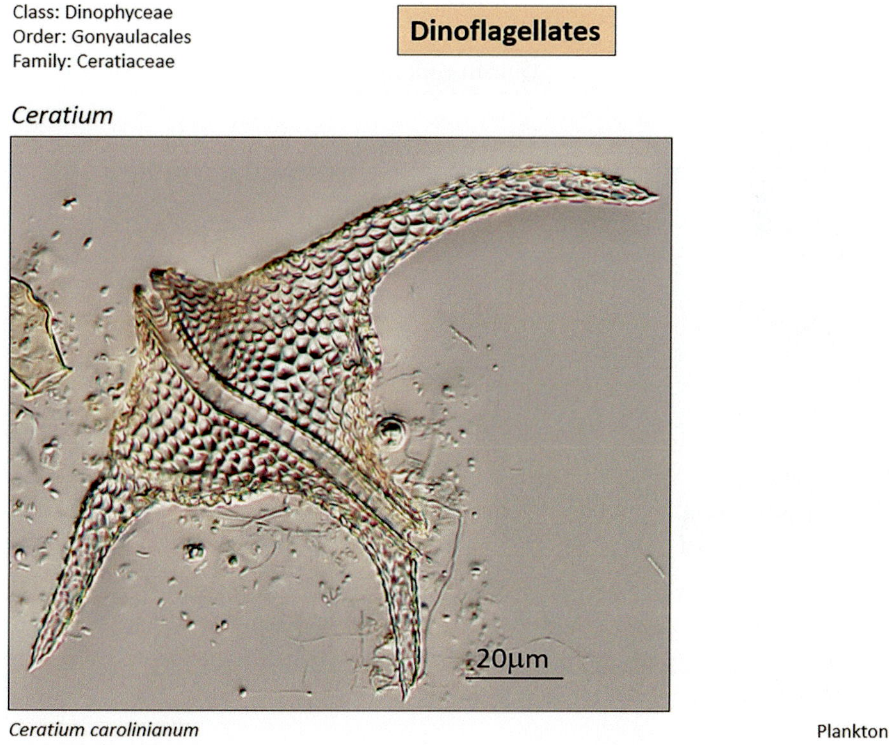

Fig. 7.4 *Nusuttodinium*

Class: Dinophyceae
Order: Gymnodiniales
Family: Gymnodiniaceae

Dinoflagellates

Nusuttodinium

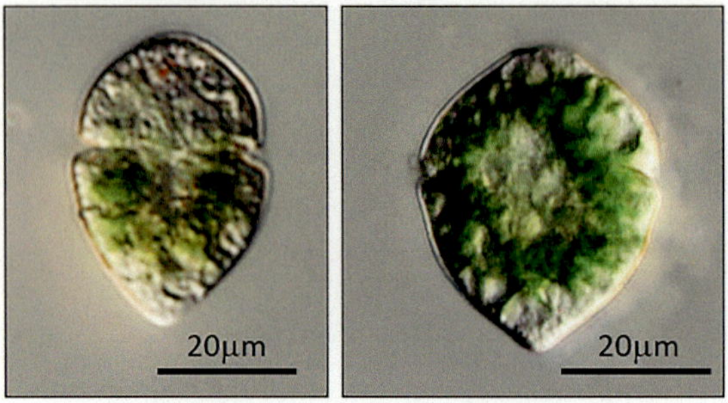

Nusuttodinium aeruginosum Plankton

Fig. 7.5 *Gymnodinium*

Class: Dinophyceae
Order: Gymnodiniales
Family: Gymnodiniaceae

Dinoflagellates

Gymnodinium

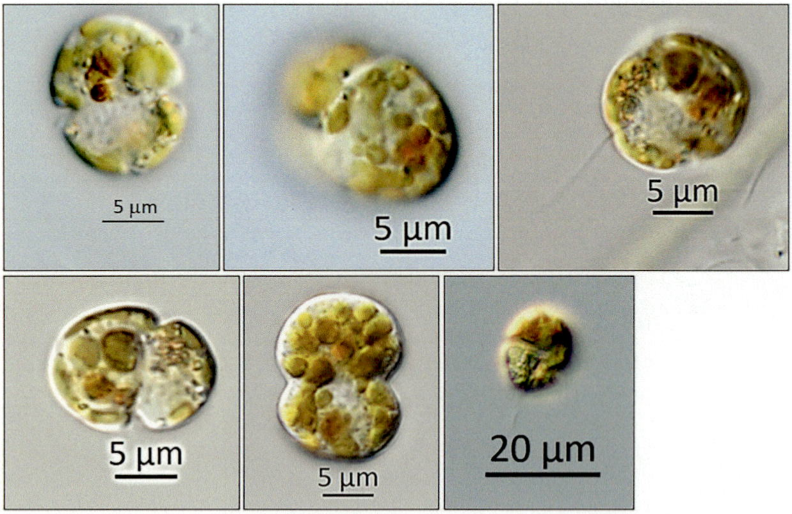

Gymnodinium luteofaba Plankton

Fig. 7.6 *Akashiwo*

Class: Dinophyceae
Order: Akashiwales
Family: Akashiwaceae

Dinoflagellates

Akashiwo

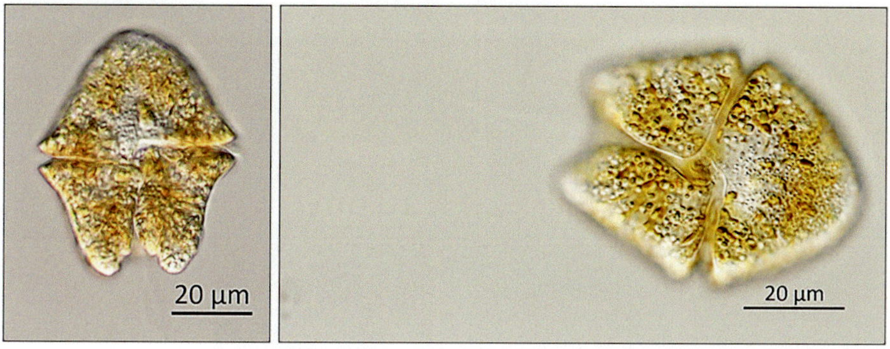

Akashiwo sanguinea Plankton Caloosahatchee River, Florida, USA

Fig. 7.7 *Karenia brevis*

Class: Dinophyceae
Order: Gymnodiniales
Family: Gymnodiniaceae

Dinoflagellates

Karenia brevis

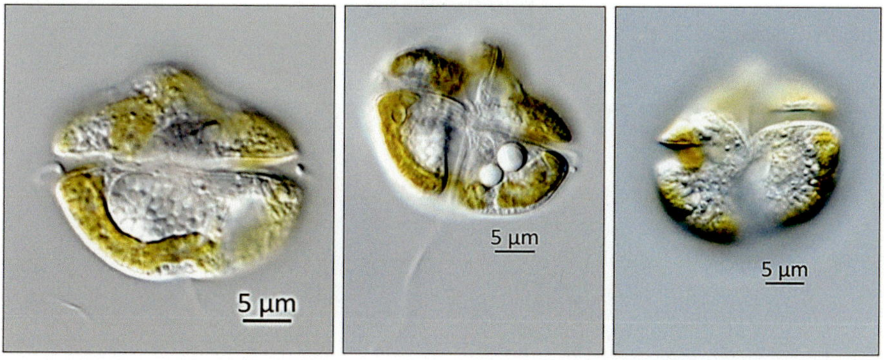

Karenia brevis Plankton red tide-former; marine

Fig. 7.8 *Parvodinium*

Class: Dinophyceae
Order: Peridiniales
Family: Peridiniopsidaceae

Dinoflagellates

Parvodinium

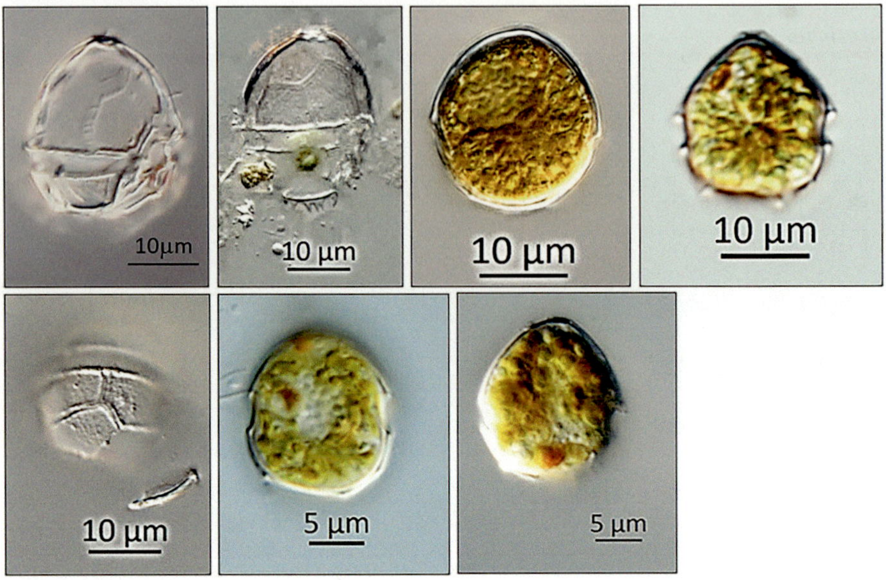

Parvodinium spp.

Fig. 7.9 *Parvodinium*

Class: Dinophyceae
Order: Peridiniales
Family: Peridiniopsidaceae

Dinoflagellates

Parvodinium (resting cysts)

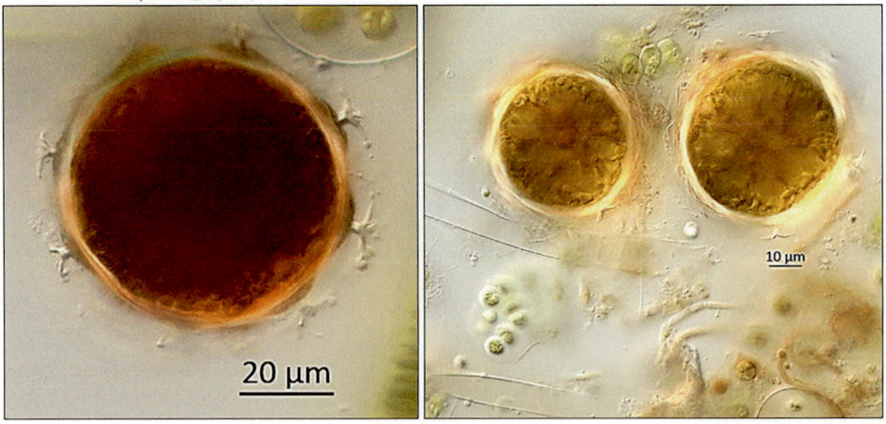

Parvodinium sp. Metaphyton

Fig. 7.10 *Parvodinium* (resting cysts)

Class: Dinophyceae
Order: Peridiniales
Family: Peridiniopsidaceae

Dinoflagellates

Peridinium

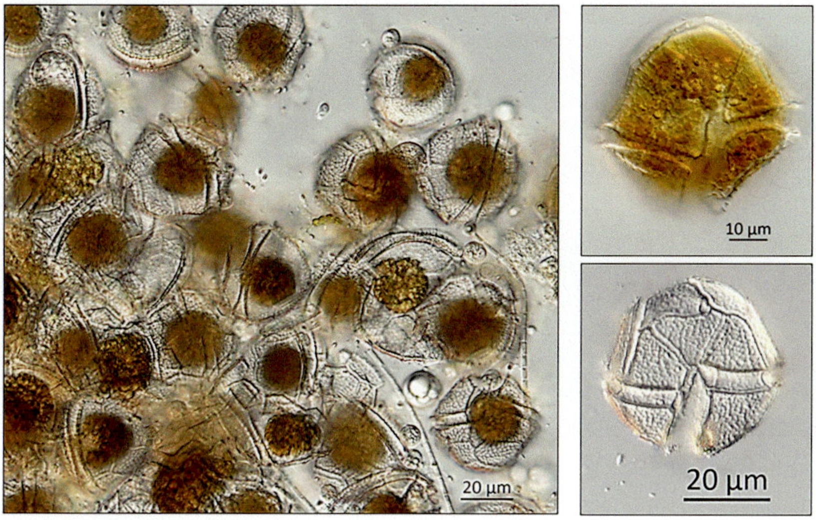

Peridinium cinctum

Plankton

Fig. 7.11 *Peridinium*

Class: Dinophyceae
Order: Peridiniales
Family: Peridiniopsidaceae

Dinoflagellates

Peridiniopsis

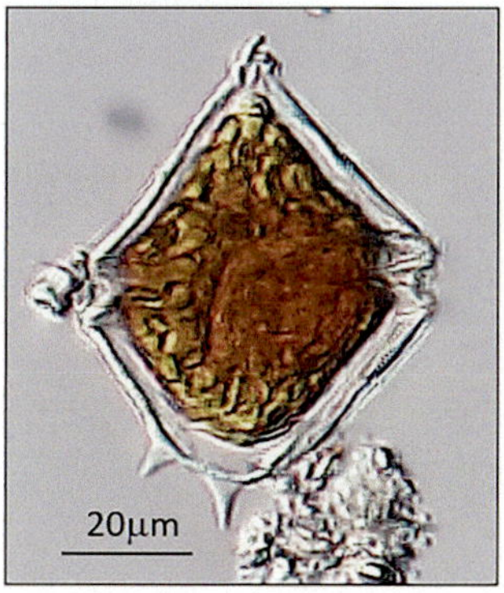

Peridiniopsis cunnigtonii

Plankton

Fig. 7.12 *Fusiperidinium*

Class: Dinophyceae
Order: Thoracosphaerales
Family: Thoracosphaeraceae

Dinoflagellates

Fusiperidinium

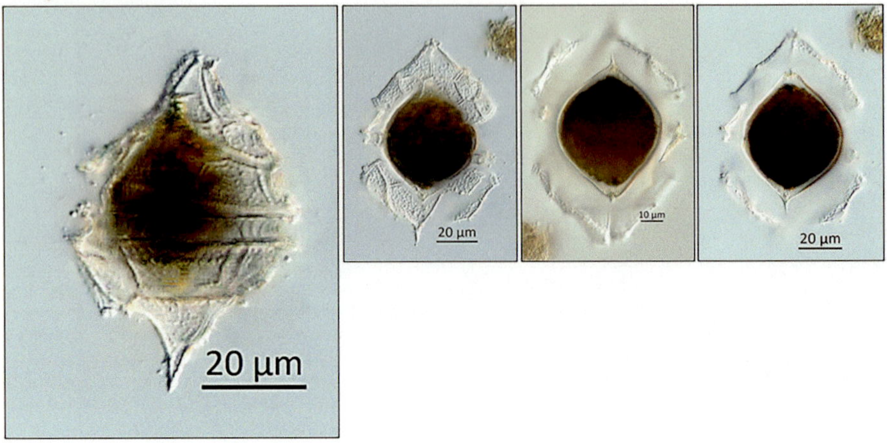

Fusiperidinium wisconsinense Plankton

Fig. 7.13 *Peridiniopsis*

Class: Dinophyceae
Order: Suessiales
Family: Lophodiniaceae

Dinoflagellates

Lophodinium

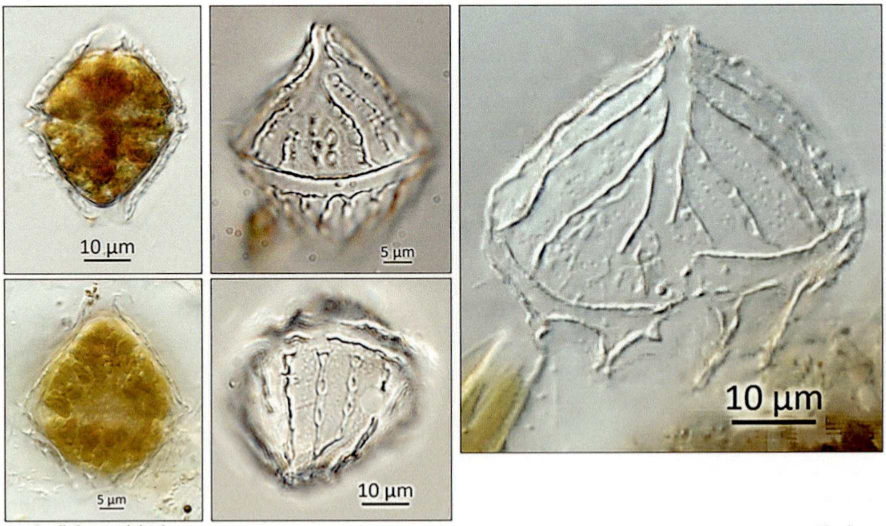

Lophodinium polylophum Plankton

Fig. 7.14 *Lophodinium*

Class: Dinophyceae
Order: Phytodiniales
Family: Phytodiniaceae

Dinoflagellates

Dinococcus

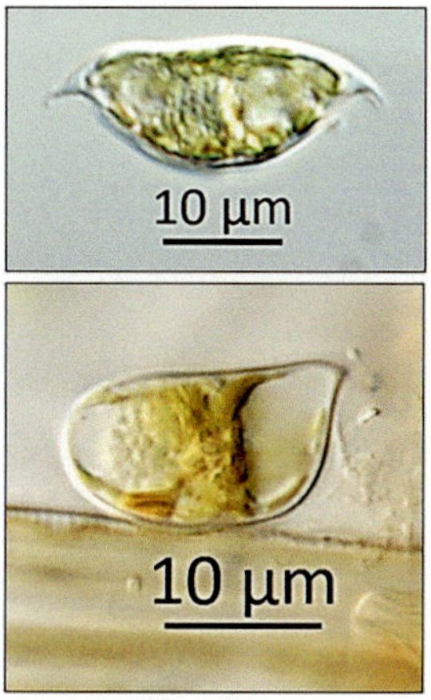

Dinococcus bicornis

Cystodinium

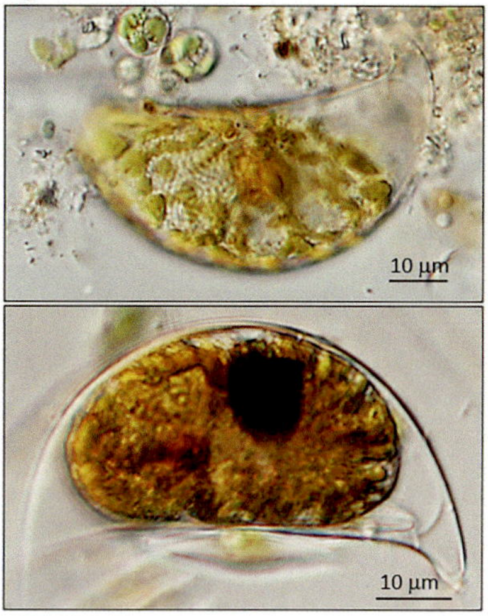

Cystodinium sp.

Epiphyton

Fig. 7.15 *Dinococcus* and *Cystodinium*

Class: Dinophyceae
Order: Phytodiniales
Family: Phytodiniaceae

Dinoflagellates

Cystodinedria

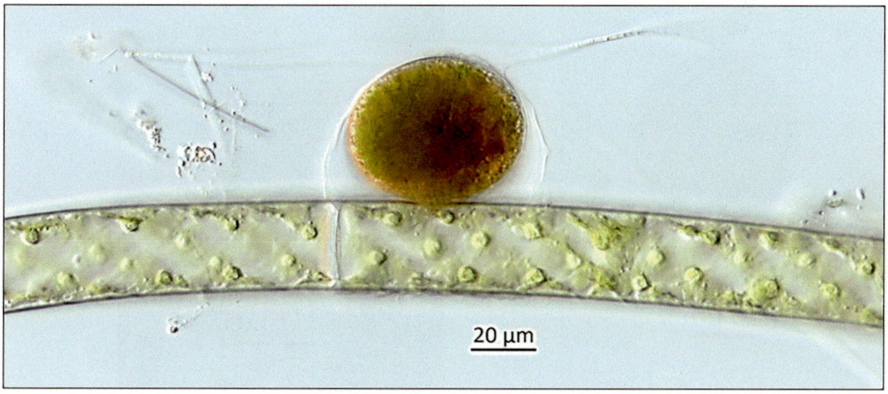

Cystodinedria sp.

Epiphyton

Fig. 7.16 *Cystodinedria*

Class: Dinophyceae
Order: Phytodiniales
Family: Phytodiniaceae

Dinoflagellates

Hypnodinium

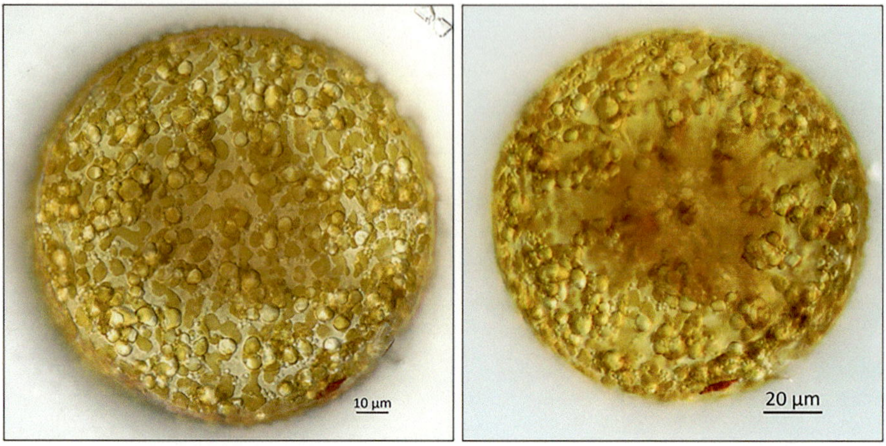

Hypnodinium sphaericum Metaphyton

Fig. 7.17 *Hypnodinium*

Class: Dinophyceae
Order: Phytodiniales
Family: Phytodiniaceae

Dinoflagellates

Stylodinium

Stylodinium globosum *Stylodinium longipes*

Epiphyton

Fig. 7.18 *Stylodinium*

Class: Dinophyceae
Order: Phytodiniales
Family: Phytodiniaceae

Dinoflagellates

Tetradinium

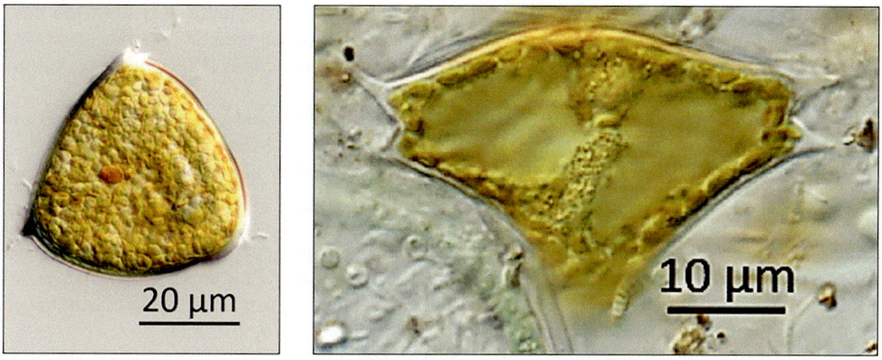

Tetradinium simplex *Tetradinium javanicum*

Epiphyton

Fig. 7.19 *Tetradinium*

Class: Dinophyceae
Order: Suessiales
Family: Hemidiniaceae

Dinoflagellates

Hemidinium

Hemidinium nasutum

Plankton

Fig. 7.20 *Hemidinium*

Suggested Reading

Carty, S. (2003). Dinoflagellates. In J. D. Wehr & R. G. Sheath (Eds.), *Freshwater algae of North America, ecology and classification* (pp. 685–714).

Carty, S. (2014). *Freshwater dinoflagellates of North America* (pp. [i]–viii, [2], [1]–260, 153 figs). Comstock Publishing Associates. A division of Cornell University Press.

Carty, S., & Cox, E. R. (1985). Observations on *Lophodinium polylophum* (Dinophyceae). *Journal of Phycology, 21*(3), 396–401.

Carty, S., & Cox, E. R. (1986). *Kansodinium* gen. nov. and *Durinskia* gen. nov.: two genera of freshwater dinoflagellates (Pyrrhophyta). *Phycologia, 25*(2), 197–204.

Cavalier-Smith, T., & Chao, E. E. (2004). Protalveolate phylogeny and systematics and the origins of Sporozoa and dinoflagellates (phylum Myzozoa nom. nov.). *European Journal of Protistology, 40*(3), 185–212.

Fawcett, R. C., & Parrow, M. W. (2012). Cytological and phylogenetic diversity in freshwater *Esoptrodinium/Bernardinium* species (Dinophyceae). *Journal of Phycology, 48*(3), 793–807.

Gómez, F. (2012). A quantitative review of the lifestyle, habitat and trophic diversity of dinoflagellates (Dinoflagellata, Alveolata). *Systematics and Biodiversity, 10*, 267–275.

Lira, B., Parrow, M. W., & Tavera, R. (2017). Morphology and ecology of freshwater-blooming *Durinskia baltica* (Dinophyceae: Peridiniales) in Xochimilco, Mexico. *Microbiology Research Journal International, 18*(1), 1–15.

8 Rhodophytes and Glaucophytes

Glaucophytes and red algae are the most primitive of the eukaryotic algae in the kingdom Plantae. The glaucophytes have primitive chloroplasts that closely resemble cyanobacteria and have peptidoglycan in their cell wall. Their blue-green pigmentation is from phycocyanin, which is organized into a phycobilisome-like cyanobacteria. This is also true for red algae, which is dominated by phycoerythrin, the red pigment. These pigments act as an antenna and allow them to absorb blue light and live in habitats where other wavelengths are less available. In both phyla, the pigments are embedded in membranes (thylakoids) that are not stacked, and cells do not have a centriole, an internal cellular structure involved in chromosome movement. Red algae never have a flagellated stage. As with all algae, red algae have chlorophyll *a* but also have chlorophyll *d* and carotenoids.

Most red algae are marine; however, a few genera are found in freshwater habitats including ponds, lakes, hot springs, and especially in clean streams and rivers and can thrive in both nutrient-rich and oligotrophic waters. Red algae store their photosynthate as Floridian starch, a semi-amylopectin without amylose, which is stored in the cell's cytoplasm (cytosol). Carbon captured during photosynthesis is stored in the cytosol of red algae, compared to chloroplast storage found in most other eukaryotic algae, and is considered an ancestral mechanism. The cell wall of red algae has cellulose and an extensive matrix of polysaccharides, making them "slimy" to touch. Epiphytes have a difficult time attaching to red algae because of the constant production of the extracellular polysaccharides. These dissolve readily compared to other algae, such as some Chlorophytes with a cellulosic cell wall.

Red algae display a diversity of morphologies, ranging from single cells (not illustrated in this text) to unbranched simple filaments (uniseriate and multiseriate forms, such as *Bangia* (Fig. 8.1) to forms that have a highly branched complex thallus. In addition, many of the red algae have a complex life cycle, with a haploid (n)

B. H. Rosen, *Color Atlas of Freshwater Algae*,
https://doi.org/10.1007/978-3-031-93122-2_8

male and female gametophytes that form spermatangia and carpogonia, respectively. The spermatia are carried by water currents because red algae lack a flagellated stage, although amoeboid movements have been observed. A diploid stage (2n), a tetrasporophyte, produces a tetrasporangia, with tetraspores (n) that have undergone meiosis, which once again produce the gametophytes. Other than the reproductive structures, these alternating generations, the tetrasporophyte and the gametophyte, are morphologically indistinguishable.

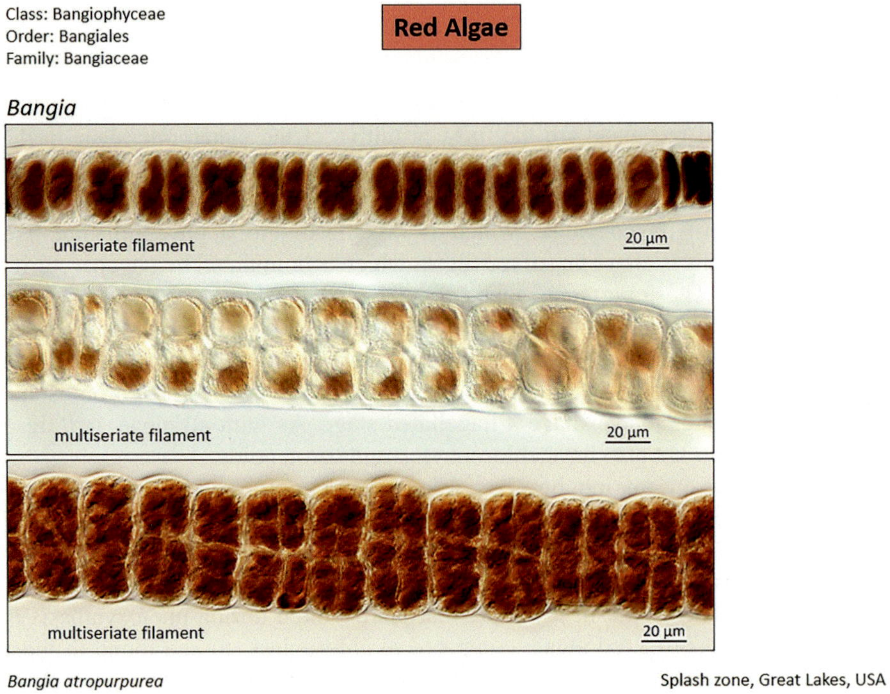

Fig. 8.1 *Bangia*

Class: Compsopogonophyceae
Order: Comsopogonales
Family: Comsopogonaceae

Red Algae

Compsopogon

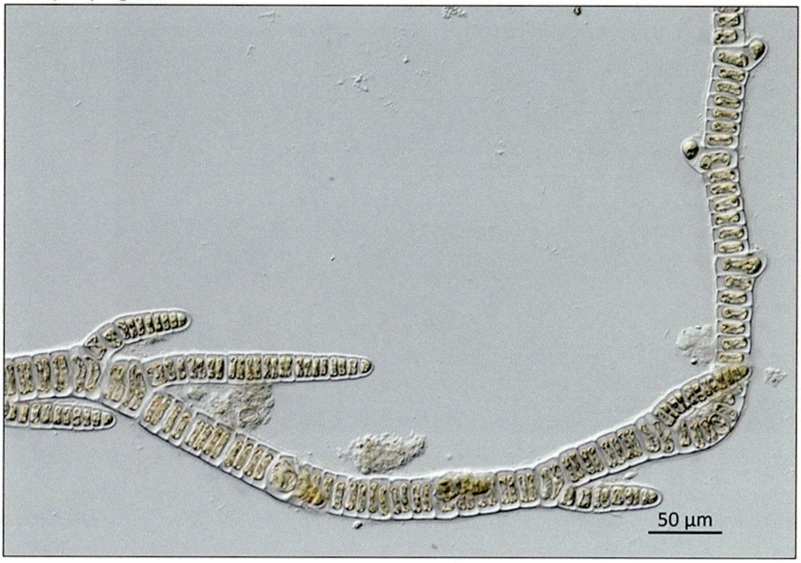

Compsopogon caeruleus Metaphyton

Fig. 8.2 *Compsopogon*

Class: Compsopogonophyceae
Order: Comsopogonales
Family: Comsopogonaceae

Red Algae

Compsopogon

Compsopogon caeruleus Metaphyton

Fig. 8.3 *Compsopogon*

Class: Compsopogonophyceae
Order: Comsopogonales
Family: Comsopogonaceae

Red Algae

Compsopogon

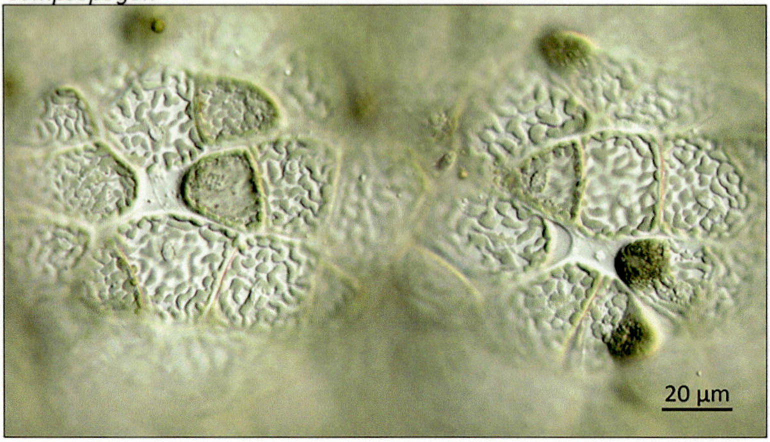

Compsopogon caeruleus Metaphyton

Fig. 8.4 *Compsopogon*

Class: Compsopogonophyceae
Order: Comsopogonales
Family: Comsopogonaceae

Red Algae

Compsopogon

Compsopogon caeruleus Metaphyton

Fig. 8.5 *Compsopogon*

Class: Florideophyceae
Order: Acrochaetiales
Family: Audouinellaceae

Red Algae

Audouinella

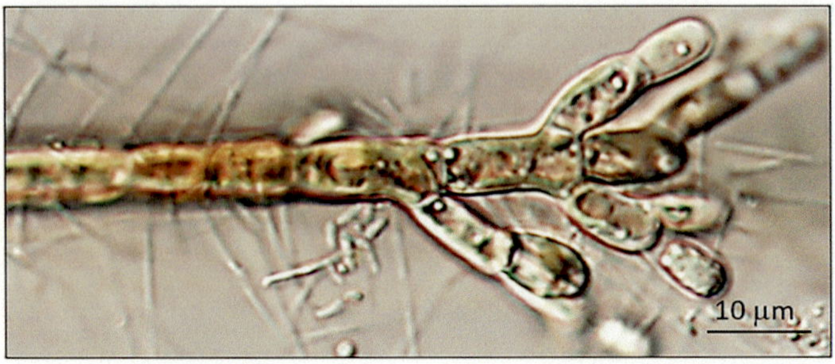

Audouinella hermannii

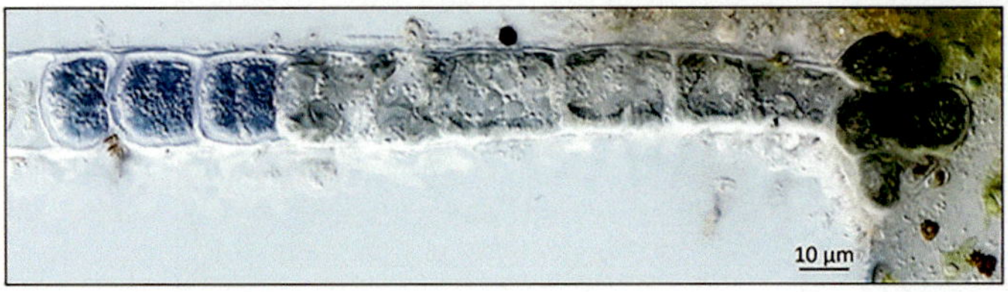

Audouinella pygmaea

Epiphyton, metaphyton

Fig. 8.6 *Audouinella*

Class: Florideophyceae
Order: Batrachospermales
Family: Batrachospermaceae

Red Algae

Batrachospermum

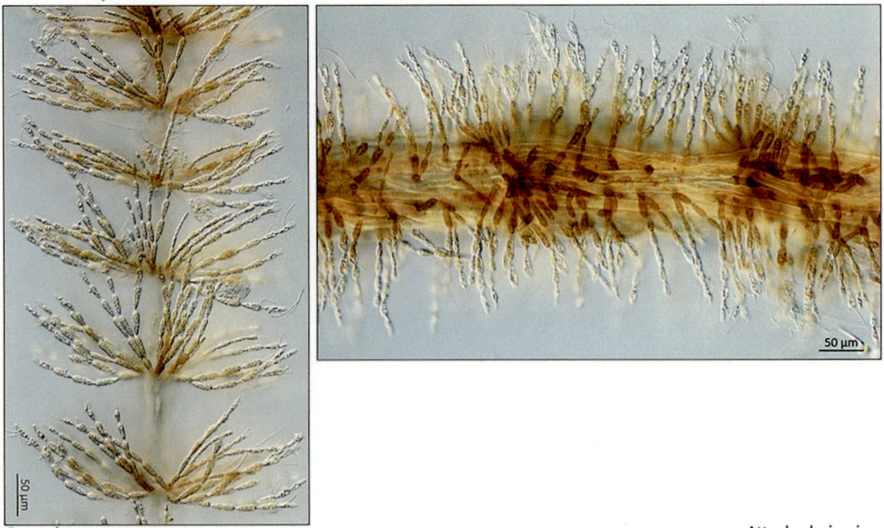

Batrachospermum sp.

Attached, riverine

Fig. 8.7 *Batrachospermum*

Class: Florideophyceae
Order: Batrachospermales
Family: Lemaneaceae

Red Algae

Lemanea

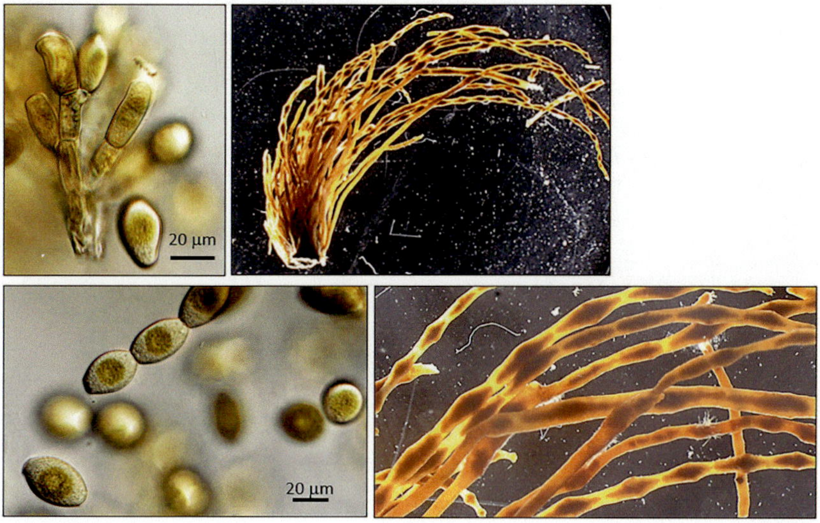

Lemanea fluviatilis

Attached, riverine

Fig. 8.8 *Lemanea*

Class: Florideophyceae
Order: Ceramiales
Family: Rhodomelaceae

Red Algae

Polysiphonia

Polysiphonia sp.

Attached; freshwater canal with marine influence, Florida, USA

Fig. 8.9 *Polysiphonia*

Class: Stylonemetophyceae
Order: Stylonematales
Family: Stylonemataceae

Red Algae

Chroodactylon

Chroodactylon ornatum Splash Zone, Lake Michigan, USA

Fig. 8.10 *Chroodactylon*

Class: Stylonemetophyceae
Order: Stylonematales
Family: Stylonemataceae

Red Algae

Chroodactylon

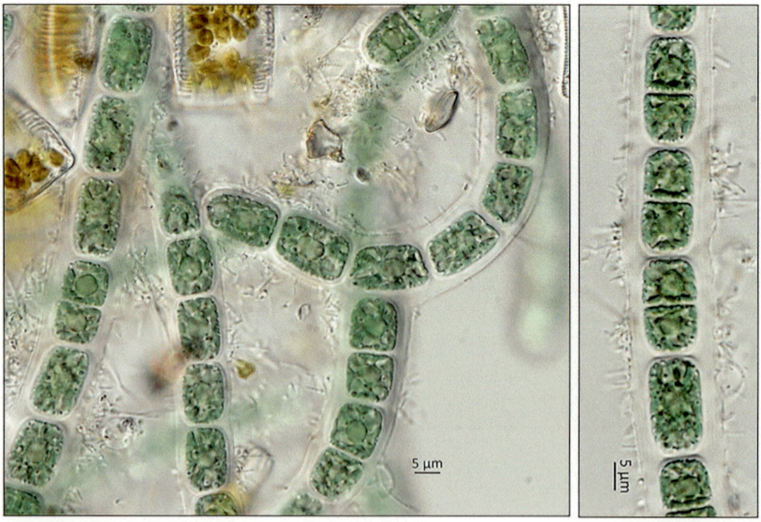

Chroodactylon ornatum Splash Zone, Lake Michigan, USA

Fig. 8.11 *Chroodactylon*

Class: Glaucophyceae
Order: Glaucocystales
Family: Glaucocystaceae

Glaucophyta

Glaucocystis

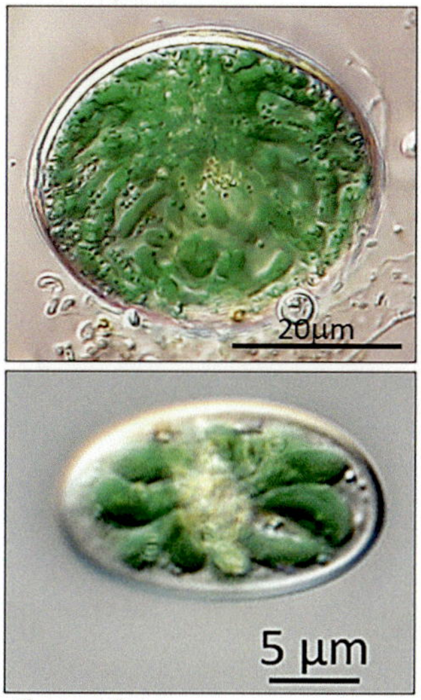

Glaucocystis spp.

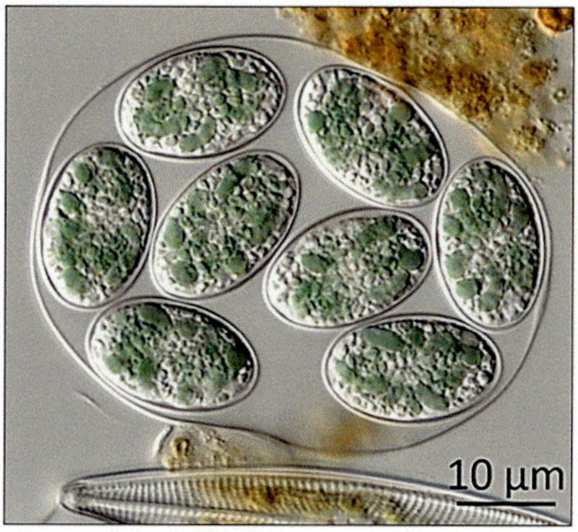

Glaucocystis nostochinearum

Metaphyton, epiphyton

Fig. 8.12 *Glaucocystis*

Class: Glaucophyceae
Order: Gloeochaetales
Family: Gloeochaetales *familia incertae sedis*

Glaucophyta

Cyanoptyche

Cyanoptyche gloeocystis

Metaphyton, epiphyton

Fig. 8.13 *Cyanoptyche*

Class: Glaucophyceae
Order: Glaucocystales
Family: Glaucocystaceae

Glaucophyta

Cyanophora paradoxa

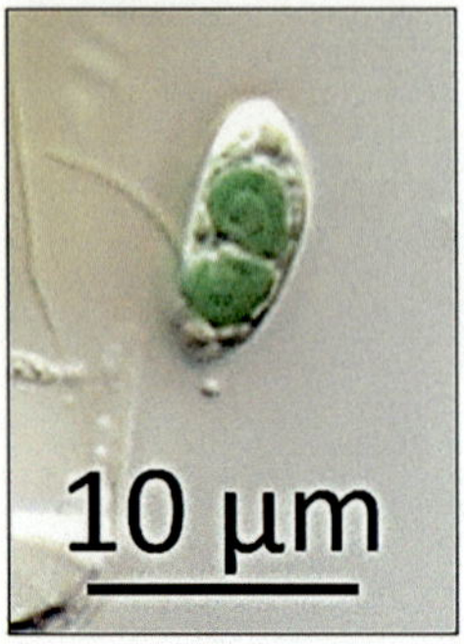

Cyanophora paradoxa Phytoplankton

Fig. 8.14 *Cyanophora paradoxa*

Suggested Reading

Cole, K. M., & Sheath, R. G. (Eds.). (1990). *Biology of the red algae* (pp. ix + 517). Cambridge University Press.

Drew, K. M., & Ross, R. (1965). Some generic names in the Bangiophycidae. *Taxon, 14*, 93–99.

Hambrook, J. A., & Sheath, R. G. (1991). Reproductive ecology of the fresh water red alga *Batrachospermum boryanum* Sirodot in a temperate headwater stream. *Hydrobiol, 218*, 233–246.

Javornicky, P. (2016). The occurrence of microscopic algae with blue-green chloroplasts or with endocyanelles (Glaucophyta) in the fresh waters of The Czech Republic, with a new report of Cryptella cyanophora Pascher. *Fottea, 16*(1), 22–33.

Korch, J. E., & Sheath, R. G. (1989). The phenology of *Audouinella violacea* (Acrochaetiaceae, Rhodophyta) in a Rhode Island stream (USA). *Phycologia, 28*, 228–236.

Korshikov, A. A. (1924). Protistologicheskiye zametki [Protistological notes]. *Russkii Arkhiv Protistologii, 3*(1–2), 57–74.

Masjuk, N. P. (1999). Cyanelles and taxonomic status of cyanomes. *International Journal on Algae, 1*(1), 3–18.

Müller, K. M., Cole, K. M., & Sheath, R. G. (2003). Systematics of *Bangia* (Bangiales, Rhodophyta) in North America. II. Biogeographical trends in karyology: Chromosome numbers and linkage with gene sequence phylogenetic trees. *Phycologia, 42*, 209–219.

Necchi, O., Jr., & Vis, M. L. (2021). Subphylum cyanidiophytina, class cyanidiophyceae; subphylum proteorhodophytina, classes compsopogonophyceae, porphyridiophyceae, rhodellophyceae, and stylonematophyceae. In *Freshwater red algae phylogeny, taxonomy and biogeography* (pp. 27–56).

Pascher, A. (1929). Studien über symbiosen. Über einige Endosymbiosen von Blaualgen in Einzellern. *Jahrbücher für wissenschaftliche Botanik, 71*, 386–462.

Rintoul, T. L., Sheath, R. G., & Vis, M. L. (1999). Systematics and biogeography of the Compsopogonales (Rhodophyta) with emphasis on the freshwater families in North America. *Phycologia, 38*, 517–527. 5 figs, 3 tables.

Sheath, R. G. (1984). The biology of freshwater red algae. *Progress in Phycological Research, 3*, 89–157.

Sheath, R. G., & Cole, K. M. (1980). Distribution and salinity adaptations of *Bangia atropurpurea* (Rhodophyta), a putative migrant into the Laurentian great lakes. *Journal of Phycology, 16*, 412–420.

Sheath, R. G., Vis, M. L., & Cole, K. M. (1993). Distribution and systematics of the freshwater red algal family Thoreaceae in North America. *European Journal of Phycology, 28*, 231–241.

Sheath, R. G., Müller, K. M., Vis, M. L., & Entwisle, T. J. (1996). A re-examination of the morphology, ultrastructure and classification of genera in the Lemaneaceae (Batrachospermales, Rhodophyta). *Phycological Research, 44*, 233–246.

Takahashi, T., Nishida, T., Tuji, A., et al. (2016). Delineation of six species of the primitive algal genus *Glaucocystis* based on *in situ* ultrastructural characteristics. *Scientific Reports, 6*, 29209. https://doi.org/10.1038/srep29209

Vis, M. L., & Necchi, O., Jr. (2021). *Freshwater red algae phylogeny, taxonomy and biogeography* (pp. [i]–xii, 1–338). Springer.

Vis, M. L., & Sheath, R. G. (1992). Systematics of the freshwater red algal family Lemanaceae in North America. *Phycologia, 31*(2), 164–179.

Vis, M. L., & Sheath, R. G. (1996). Distribution and systematics of *Batrachospermum* (Batrachospermales, Rhodophyta) in North America. 9. Section *Batrachospermum* : Description of five new species. *Phycologia, 35*, 124–134.

Vis, M. L., Sheath, R. G., Hambrook, J. A., & Cole, K. M. (1994). Stream macroalgae of the Hawaiian Islands : A preliminary study. *Pacific Science, 48*(2), 175–187.

Vis, M. L., Hodge, J. C., & Necchi, O., Jr. (2008). Phylogeography of *Batrachospermum macrosporum* (Batrachospermales, Rhodophyta) from North and South America. *Journal of Phycology, 44*, 882–888.

Whitton, B. A. (2011). Phylum Glaucophyta. In D. M. John, B. A. Whitton, & A. J. Brook (Eds.), *The freshwater algal flora of the British Isles* (An identification guide to freshwater and terrestrial algae) (2nd ed., pp. 766–767). Cambridge University Press.

Yoon, H. S., Muller, K. M., Sheath, R. G., Ott, F. D., & Bhattacharya, D. (2006). Defining the major lineages of red algae (Rhodophyta). *Journal of Phycology, 42*, 482–492.

MIX
Papier aus verantwortungsvollen Quellen
Paper from responsible sources
FSC® C105338

If you have any concerns about our products, you can contact us on
ProductSafety@springernature.com

In case Publisher is established outside the EU, the EU authorized representative is:
Springer Nature Customer Service Center GmbH
Europaplatz 3, 69115 Heidelberg, Germany

Printed by Libri Plureos GmbH
in Hamburg, Germany